In the Loyal
Mountains

In the Loyal Mountains

RICK BASS

A MARINER BOOK

HOUGHTON MIFFLIN COMPANY

BOSTON   NEW YORK

For information about permission to reproduce selections from this book, write to Permissions, Houghton Mifflin Company, 215 Park Avenue South, New York, New York 10003.

Library of Congress Cataloging-in-Publication Data
Bass, Rick, date.
    In the Loyal Mountains / Rick Bass.
        p.   cm.
    Contents: The history of Rodney — Swamp boy — Fires — The valley — Antlers — Wejumpka — The legend of Pig-eye — The wait — Days of heaven — In the Loyal Mountains.
    ISBN 0-395-71687-X   ISBN 0-395-87747-4 (pbk.)
    I. Title.
PS3552.A821315   1995
813'.54 — dc20   94-49609
                    CIP
Printed in the United States of America

QUM 10 9 8 7 6 5 4 3 2 1

Acknowledgments
These stories first appeared, in different versions, in the following publications: "The History of Rodney" in *Ploughshares;* "Swamp Boy" in *Beloit Fiction Journal;* "Fires" in *The Quarterly* and *Big Sky Journal;* "The Valley" in *American Short Fiction;* "Antlers" in *Special Report;* "Wejumpka" in *The Chariton Review;* "The Legend of Pig-Eye" in *The Paris Review;* "The Wait" in *Story;* "Days of Heaven" in *Ploughshares;* "In the Loyal Mountains" in *Southwest Review.* These stories were also published in the following anthologies: "The History of Rodney" in *New Stories from the South: The Year's Best 1990* and *Other Sides of Silence: New Fiction from Ploughshares;* "Fires" in *Pocketful of Prose;* "The Valley" in *Listening to Ourselves;* "Antlers" in *Texas Bound, In the Company of Animals,* and *Best of the West 4;* "Wejumpka" in *Pushcart Prize XV* and *New American Short Stories 2;* "The Legend of Pig-Eye" in *The Best American Short Stories 1991;* "The Wait" in *Boats;* "Days of Heaven" in *Pushcart Prize XVIII* and *The Best American Short Stories 1992;* "In the Loyal Mountains" in *New Stories from the South: The Year's Best 1991.* Grateful acknowledgment is made to these publications and their editors.

    Deepest thanks to Russell Chatham for the painting on the jacket, to Hilary Liftin for production assistance, to Melodie Wertelet for the book's design, and most especially to my editors, Camille Hykes and Larry Cooper. My publisher Sam Lawrence, who passed away last year, was many things to many people, but for everyone he was a steadfast lover of books. He is missed.

    These stories are products of the imagination. The characters in them are not intended to represent any real persons.

FOR
JOHN GRAVES
JIM HARRISON
TOM McGUANE

CONTENTS

The History
of Rodney

IT RAINS in Rodney in the winter. But we have history; even for Mississippi, we have that. Out front there's a sweet olive tree that grows all the way up to the third story where Elizabeth's sun porch is. Through the summer butterflies swarm in the front yard, drunk on the smell of the tree. But in the winter it rains.

The other people in the town of Rodney are the daughters, sons, and granddaughters of slaves. Sixteen thousand people lived in Rodney before and during the Civil War. Now there are a dozen of us.

This old house I rent costs fifty dollars a month. Electricity sizzles and arcs from the fuse box on the back porch and tumbles to the ground in bouncing blue sparks. The house has thirty-five rooms, some of which are rotting — one has a tree growing through the floor — and the ceilings are all high, though not as high as the trees outside.

Here in the ghost town of Rodney there is a pig, a

murderer, that lives under my house, and she has killed several dogs. The pig had twenty piglets this winter, and like the bad toughs in a western, they own the town. When we hear or see them coming, we run. We could shoot them down in the middle of the dusty lane that used to be a street, but we don't: we're waiting for them to fatten up on their mother's milk.

We're also waiting for Preacher to come back. He's Daisy's boyfriend, and he's been gone for forty years.

Back in the trees, loose peafowl scream in the night. It is like the jungle out there. The river that used to run past Rodney — the Mississippi, almost a mile wide — shifted course exactly one hundred years ago.

It happened overnight. The earthen bulge of an oxbow, a bend upstream, was torn by the force of the water. Instead of making its taken-for-granted way through the swamp — the slow wind of northern water down from Minnesota — the river pressed, like sex, and broke through.

I've been reading about this in the old newspapers. And Daisy, who lives across the street, has been telling me about it. She says that the first day after it happened, the townspeople could do nothing but blink and gape at the wide sea of mud. Rodney then was the second-largest port in the South, second only to New Orleans.

Boats full of cotton were stranded in the flats. Alligators and snakes wriggled in the deep brown as the townspeople waited for a rain to come and fill the big river back up. Giant turtles crept through the mud and moved on, but the great fish could do nothing but die. Anchors and massive logs lay strewn on the river bottom. Birds gathered overhead and circled the dying fish cautiously, now and then landing in the fetid mud. When the fish began to smell bad, the people in Rodney packed what belongings they could and hiked into the bluffs and jungle above the river to escape the rot and disease.

When the mud had dried and grown over with lush tall grass, the townspeople moved back. Some of the men tracked the river, hunting it as if it were a wounded animal, and they found it seven miles away, running big and strong, as wide as it had ever been. It was flowing like a person's heart. It had only shifted.

Daisy didn't see the river leave, but her mother did. Daisy says that the pigs in Rodney are descended from Union soldiers. The townspeople marched the soldiers into the Presbyterian church one Sunday, boarded up the doors and windows, and then Daisy's mother turned them all into pigs.

The mother pig is the size of a small Volkswagen; her babies are the color and shape of footballs. They grunt and snort at night beneath Elizabeth's and my house.

Daisy has a TV antenna rising a hundred and fifty feet into the air, above the trees. Daisy can cure thrash, tuberculosis, snakebite, ulcers, anything as long as it does not affect someone she loves. She's powerless then; she told me so. She cooks sometimes for Elizabeth and me. We buy the food and give her some money and she cooks: fried eggs, chicken, okra. Sometimes Elizabeth isn't hungry — she'll be lying on the bed up in the sun room, wearing just her underpants and sunglasses, reading a book — so I'll go over to Daisy's by myself.

We live so far from civilization. The mail comes only once a week, from Natchez. The mailman is frightened of the pigs. Sometimes they chase his jeep up the steep hill, up the gravel road that leads out of town. Their squeals of rage are a high, mad sound, and they run out of breath easily.

Daisy never gets mail. We let her come over and read ours.

"This used to be a big town," she said when she came over to introduce herself. She gestured out to the cotton field behind her house. "A port town. The river used to lay right out there."

"Why did it leave?" Elizabeth asked.

Daisy shook her head and wouldn't answer.

"Will you take us to the river?" I asked. "Will you show it to us?"

Daisy shook her head again. "Nope," she said, drawing circles in the dust with her toe. "You got to be in *love* to see the river," she said, looking at me and then at Elizabeth.

"Oh, but we are," Elizabeth cried, taking my arm. "That's why we're here!"

"Well," Daisy said. "Maybe."

Daisy likes to tell us about Preacher; she talks about him all the time. He was twenty, she was nineteen. Once there was a Confederate gunboat in the cotton field. The boat has since rusted away to nothing, but it was still in fair shape when Preacher and Daisy lived on it, out in the middle of the field, still rich and growing green with cotton, the color of which is heat-hazy in the fall. They slept in the captain's quarters on a striped mattress with no sheets. They rubbed vanilla on their bodies to keep the bugs from biting.

There were skeletons in the boat and in the field, skeletons of sailors who had drowned when the ship burned and sank from cannon fire to the bow. But these were old bones and no more harmful than, say, a cow's skull, or a horse's.

She and Preacher made love on the tilted deck, Daisy said, through the blazing afternoons. Small breezes cooled them. They made love at night, too, with coal-oil lamps burning around the gunwales. Their cries were so loud, she said, that birds roosting in the swamp took flight into the darkness and circled overhead.

"All we were going to do was live out on that boat and make love mostly all day," she said. "Preacher wasn't hurtin' anybody. We had a garden, and we went fishing. We skimmed the river in our wood canoe. One day he caught a porpoise. It had come all the way up from the gulf after a rainstorm and was confused by the fresh water. It pulled us all over the river for a whole afternoon."

A whole afternoon. I could see the porpoise leaping, and I could see Daisy as she was then, with a straw hat low around her brow. I could see Preacher leaning forward, battling the big fish.

"It got away," Daisy said. "It broke the line." She was sitting on the porch, shelling peas from her garden, remembering. "Oh, we both cried," she said. "Oh, we wanted that fish."

Elizabeth and I live here quietly, smoothing things over, making the country tame again; but it is like walking on ice. Sometimes I imagine I can hear echoes, noises and sounds from a long, long time ago.

"This place isn't on the map, right?" Elizabeth will ask. It's a game we play. We're frightened of cities, of other people.

"It might as well not even exist," I'll tell her.

She seems reassured.

The seasons mix and swirl. Except for the winter rains and the hard, stifling brutality of August, it's easy to confuse one season with the next. Sometimes wild turkeys gobble and fan in the dust of the road, courting. Their lusty gobbles awaken us at daylight — a watery, rushing sound. That means it is April, and the floodwaters will not be coming back. Every sight, every small scent and sound, lies still, its own thing, as if there are no seasons. As if there is only one season.

I'm glad Elizabeth and I have found this place. We have not done well in other places. Cities — we can't understand them. In a city everything seems as though it is over so fast: minutes, hours, days, lives.

Daisy keeps her yard very neat; she cuts it with the push mower weekly. Tulips and roses line the edges of it. She's got two little beagle pups, and they roll and wrestle in the front yard and on her porch. Daisy conducts church services in the abandoned Mount Zion Baptist Church, and sometimes we go. Daisy's a good preacher. The church used to be on the river bank, but now it looks out on a cotton field.

Daisy's sister, Maggie, lives in Rodney too. She used to have a crush on Preacher when he was a little boy. She says he used to sleep, curled up in a blanket, in a big empty cardboard box at the top of a long playground slide in front of the church. The slide is still there, beneath some pecan trees. It's a magnificent slide, the kind you find in big city parks. Mostly wood, it's tall and steep, rubbed shiny and smooth. It's got a little cabin or booth at the top, and that's where Preacher used to sleep, Maggie says. He didn't have any parents.

The cabin kept the rain off. Sometimes the two girls sat up there with him and played cards. They'd take turns sliding down in the cardboard box, and they'd watch as white chickens walked past, pecking at the dust and clucking.

"Maybe he always wanted one for a pet," Maggie says, trying to figure it out.

Daisy says they can put you away in Whitfield for forty years for being a chicken chaser. That was what the social workers saw when they happened to pass through Rodney once: Preacher chasing chickens down the street like a crazy man. He was just doing it for fun — well, he might have been a little hungry, Daisy owns up — but they took him away.

Daisy says she's been keeping track on her calendars. There are old ones tacked to every wall in her house, beginning with the year they hauled Preacher away. The forty years will be up this fall. She expects he'll be back after that; he'll be coming back any day.

Forty years — all for maybe what was just a mistake. Maybe he only wanted one for a pet.

Toward dark the mother pig lures dogs into the swamp. She runs down the middle of what used to be Main Street in a funny high-backed hobble, as if she's wounded, with all the little runt piglets running ahead of her, protected. A foolish dog follows, slavering at the thought of fresh and easy meat.

When the pig reaches the woods, she disappears into the heavy leafiness and undergrowth, and the dog goes in after her. Then we hear the squawls and yelps of the dog being killed.

The sow sometimes kills dogs in the middle of the day. She simply tramples them, the way a horse would. I'd say she weighs about six or seven hundred pounds, maybe more. Elizabeth carries a rifle when we go for our walks, an old seven-millimeter Mauser, slung over her shoulder on a sling, a relic from the First World War, which never affected Rodney. If that pig charged us, it would rue the day. Elizabeth is a crack shot.

"Are the pigs really cursed people?" Elizabeth asks one evening. We're over on Daisy's porch. Maggie is there too, shelling peas. Fireflies are blinking, floating out in the field as if searching with lanterns for something.

"Oh my, yes," Maggie says. "That big one is a general."

"I want to see the river," Elizabeth says for the one hundredth time, and Daisy and Maggie laugh.

Daisy leans forward and jabs Elizabeth's leg. "How you know there even *is* a river?" she asks. "How you know we're not foolin' you?"

"I can smell it," Elizabeth says. She places her hand over her heart and closes her eyes. "I can *feel* it."

Elizabeth and I put fireflies in empty mayonnaise jars, screw the lids on tight, and punch holes in the tops. We decorate our porch with them at night, or we line the bed with them, and then laugh as we love, with their blinking green bellies going on and off like soft, harmless firecrackers, as if they are applauding. It's as though we have become Preacher and Daisy. The firefly bottle-lights seem like the coal-oil lamps that lined the sides of their boat in the field. Sometimes we, too, shout out into the night.

The bed: buying it for this old house is one of the best things we've ever done. It's huge, a four-poster, and looks as if it came straight off the set of *The Bride of Frankenstein.* It has a lace canopy and is sturdy enough to weather our shaking. We have to climb three wooden steps to get into it, and sleeping in it is like going off on some final voyage, so deep is our slumber, so quiet are the woods around us and what is left of the town. The birds will not scream until further into the night, so they are part of our dreams, but comforting. Nothing troubles our sleep, nothing.

Before falling into that exhausted, peaceful sleep, I slip over to the window and unscrew the lids of the jars, releasing the groggy, oxygen-deprived fireflies into the fresh night air. I shake each bottle to make sure I get them all out. They float feebly down into the bushes, blinking wanly. Wounded paratroopers, they return to their world, looking for something, searching for the world they own and know. If you keep them in a bottle too long, they won't blink anymore.

Elizabeth loves to read. She has books stacked on all the shelves in her sun room, and in all corners, books rise stacked to the ceiling. Sometimes I take iced tea up to her in a pitcher, with lemons and sugar. I don't go in with it; I just peek through the keyhole. When she's not near naked from the heat, she sometimes puts on a white dress with lace as she reads. Her hair's dark, but there by the window it looks washed with light, and she becomes someone entirely different. She disappears when she goes into those books. She disappears and that strange, solitary light seals and bathes her escape. I knock on the door to let her know the iced tea is ready whenever she feels like getting it. Then I go back down the stairs.

After a while I hear the click of the door and the scrape of the tray as she picks it up and carries it to her table. She shuts the door with the back of her foot, or so I imagine, before she goes back to reading, holding the book in one hand while fanning herself with a little cardboard fan in the other hand. I'll go out and sit on the front steps and picture her drinking the tea, and think that I can taste its coolness and freshness.

In summer, even beneath the sweet olive tree, I sweat, but not the way she does in the oven of her upstairs room. There's no air conditioner, no ceiling fan, and late in the afternoon each day, when she takes off the white lacy dress, it is soaking wet. She rinses it out in the sink and hangs it on the porch to dry in the night breezes. The dress smells of sweet olive the next morning.

"We were going to have a baby," Daisy says. "We were

just about ready to start when they took him away. We were going to start that week so the baby would be born in the summer."

The slow summer. The time when nothing moves forward, when everything pauses, and then stops. It's a good idea.

In August men come from all over this part of the state to pick cotton. The men pick by hand. They do not leave much behind. It's like a circus. Old white horses appear — perhaps they belong to the pickers, though perhaps they come from somewhere deep in the woods — and they stand out in the cotton and watch. Behind them, red tractor-trailers rise against blue sky, and behind them, the trees. Behind is the river, which we cannot see but have been told is there.

Then the men are gone, almost all of the cotton is gone, and there are leaves on the roofs of the houses, leaves in our yards. They're brown and dried up and curled, and the street is covered with them like a carpet. The sweet olive tree doesn't lose its leaves, but the other trees do. You can hear the pigs rustling in the leaves at night, snuffling for acorns. In the daytime you can hear Daisy moving through them, her slow, heavy steps and the crunching of those leaves. Daisy and Maggie burn their leaves in wire baskets by the side of the old road.

Something about the fall makes us want to go to Daisy's church services. They last about thirty minutes, and mostly she recites Bible verses, sometimes making a few up, but they sound right. Then she sings for a while. She's got a good voice.

We sing too. In the early fall when everything is changing, the air takes on a stillness, and we feel like singing to liven things up.

They're old slave songs that Daisy sings, and you just hum and sway. You can close your eyes and forget about leaving the town of Rodney.

The owl calls at night. He's big and lives in our attic. There's a hole in the bedroom ceiling, and we can hear him scrabbling around at ten or eleven o'clock each night. When the moon gets full, he emerges from the trees, flies through one of our open windows, and lights the rooms. We hear him claw-scratch out to the banister, and with a grunt he launches himself. We hear the flapping of wings and then silence. He makes no sound as he flies.

He zooms through the house — third floor, second floor, first — looking for mice. He shrieks when he spots one. He nearly always catches them.

We have watched him from the corner in the big kitchen. Elizabeth was frightened at first, but she isn't now. The longer we live down here, the less frightened she is of anything. She is growing braver with age, as if bravery is a thing she will be needing more of.

Elizabeth and I want to build something that won't go away. We're not sure how to go about it, but some nights we run naked in the moonlight. We catch the old white plugs, the horses, as they wander loose in the cotton fields. We ride them across the fields toward where we think the river is, riding through the fog amid the tracings, the language, of the fireflies. But when we get down into the swamp, we get turned around, lost, and we have to turn back.

Daisy's standing out on her porch sometimes when we come galloping back. "You can't go to it," she says, laughing in the night. "It's got to come to you!"

Afternoons, in the fall, we pick up pecans in front of

the old church. We fix grilled cheese sandwiches for supper and share a bottle of wine. We sit on the porch in the frayed wicker swing and watch the moon crest the trees, watch it slide across and over the sky, and we can hear Maggie down the street, humming, weeding in her garden.

One night I awaken from sleep and Elizabeth isn't in bed with me. I look all over the house for her, and a slight, illogical panic grows as I move through each empty room.

The moon is up and everything is bathed in hard silver. She is sitting on the back porch in her white dress, which is still damp from the wash. She's barefoot, with her feet hanging over the edge. She's swinging them back and forth. In the yard the pigs are feeding, the huge mother and the little ones, small dirigibles now, all around her. There's a warm wind blowing from the south. I can taste the salt in it from the coast. Elizabeth has a book in her lap; she's reading by the light of the moon.

A dog barks a long way off, and I feel that I should not be watching. So I climb back up the stairs and get into the big bed and try to sleep.

But I want to hold on to something.

Luther, an old blind man who lived down the street, and whom we hardly ever saw, has passed on. Elizabeth and I are the only ones with backs strong enough to dig the grave; we bury him in the cemetery on the bluff. There are toppled gravestones from the 1850s, gravestones from the war with *C.S.A.* cut in the stone. Some of the people buried there were named Emancipation — it was a common name then.

We dig the hole without much trouble. It's soft, rich earth. Daisy says words as we fill the hole back up. That rain of earth, shovels of it, covering the box with him in it. I had to build the coffin out of old lumber. Sometimes in the spring, and in the fall too, rattlesnakes come out of the cane and lie on the flat gravestones for warmth. Because of the snakes, hardly anyone goes to the cemetery now.

A few years ago one of Daisy and Maggie's half-sisters died and they buried her up here. Her grave has since grown over with brambles and vines. Now there is the skeleton of a deer impaled high on the iron spikes of the fence that surrounds the graveyard. Dogs, maybe, had been chasing it, and the deer had tried to leap over the high fence. The skull seems to be opening its mouth in a scream.

I remember that we piled stones over the mound of fresh earth to keep the pigs from rooting.

Daisy has a salve, made from some sort of root, that she smears over her eyelids at night. It's supposed to help her fading vision, maybe even bring it back, I don't know. Whenever she comes over to read our mail, she just holds the letters and runs her hand over them, doesn't really look at the words. I think she imagines what each letter is saying, what history lies behind it, what chain of circumstances.

Daisy and Preacher used to go up onto the bluffs overlooking Rodney and walk among the trees. Then they would climb one of the tallest trees so they could see the river. They'd sit on a branch, Daisy says, and have a picnic. They'd feel the river breezes and the tree swaying beneath them. They would watch the faraway river for the longest time.

Afterward they would climb down and move through the woods some more, looking for old battle things — rusted rifles, bayonets, canteens. They would sell these relics to the museum in Jackson for a dollar apiece, boxing them up for the mailman, tying the boxes shut with twine, and sending them COD. There was always enough money to get by on.

Those nights that they came down off the bluff, they might go out onto their tilted ship, out in the deep river grass, or they might go to the river itself and swim. Or sometimes they would sit on a sandbar and look up, listening to the sounds of the water. Once in a while a barge would go by. In the night, in the dark, its silhouette would look like a huge gunboat.

Wild grapes grew along the riverbank, tart purple grapes, cool in the night, and they would pick and eat those as they watched the river.

"It can go just like that," Daisy says, snapping her fingers. "It can go that fast."

The pigs are growing fat. They're not piglets anymore; they are pigs. One morning a shot awakens us, and we sit up and look out the window.

Daisy is straddling one of the pigs and gutting it with a huge knife. She pulls out the pig's entrails and feeds them to her beagles. The other pigs have run off into the woods, but they will be back. It's a cool morning, almost cold, and steam is rising from the pig's open chest. Later, in the afternoon, there's the good smell of fresh meat cooking.

Maggie shoots a pig at dusk, for herself, and two of the old men from the other side of town get theirs the next day.

"I don't want any," Elizabeth says after she's slid down the banister. Her eyes are magic, she's shivering and holding herself, dancing up and down, goose-pimply. She's happy to be so young. "I feel as if I'll be jinxed," she says. "I mean, those pigs lived under our house."

The smell of pork, of frying bacon, hangs heavy over the town, like the blue haze from cannon fire.

Coyotes at night, and the peafowl screaming. Pecans underfoot. A full moon and the gleam of night cotton.

We mount an old white plug and ride in the cotton field. It's mostly stripped, ragged and picked over, forgotten-looking. Only a few stray white bolls remain, perfect snowy blossoms, untouched by the pickers. The scraggly bushes scrape against the bottoms of our bare feet as we ride. It has been four years we've been down here in the town of Rodney.

"I'm happy," Elizabeth says, squeezing my arm from behind on the horse. She taps the horse's ribs with her heels. The smell of woodsmoke, and overhead the nasal, far-off cries of geese going south. The horse plods along in the dust.

Daisy will be starting her church services again, and once more we'll be going. I'm glad Elizabeth's happy living in such a little one-horse shell of an ex-town. We'll hold hands, carry our Bible, and walk slowly down the road to Daisy's church. There will be a few lazy fireflies at dusk. We'll go in and sit on the bench and listen to Daisy rant and howl.

Then the moaning songs will start, the ones about being slaves. I'll shut my eyes and sway, and try not to think of other places and moving on.

And then this is what Elizabeth might say: "You'd better love me" — an order, an ultimatum. But she'll be teasing, playful, as she knows that's near all we can do here.

The air will be stuffy and warm in the little church, and for a moment we might feel dizzy, lightheaded, but the songs are what's important, what matters. The songs about being slaves.

There aren't any words. You just close your eyes and sway.

Maybe Preacher isn't coming, and maybe the river isn't coming back either. But I do not say these things to Daisy when she sits on the porch and waits for him. She remembers being our age and in love with him. She remembers all the things they did, so much time they had, all that time.

She throws bread crumbs in the middle of the road and stakes the white chickens inside the yard for him. We try to learn from her every day. We still have some time left.

The days go by. I think that we will have just exactly enough time to build what Elizabeth and I want to build — to make a thing that will last, and will not leave.

Swamp Boy

THERE WAS A KID we used to beat up in elementary school. We called him Swamp Boy. I say we, though I never threw any punches myself. And I never kicked him either, or broke his glasses, but stood around and watched, so it amounted to the same thing. A brown-haired fat boy who wore bright striped shirts. He had no friends.

I was lucky enough to have friends. I was unexceptional. I did not stand out.

We'd spy on Swamp Boy. We'd trail him home from school. Those times we jumped him — or rather, when those other boys jumped him — the first thing they struck was always his horn-rim glasses. I don't know why the thick, foggy-lensed glasses infuriated them so much. Maybe they believed he could see things with them, invisible things that they could not. This possibility, along with some odd chemistry, seemed to drive the boys into a frenzy. We would go after him into the old woods along the bayou that he loved. He went there every day.

We followed him out of school and down the winding clay road. The road led past big pines and oaks, past puddles of red water and Christ-crown brambles of dewberries, their white blossoms floating above thorns. He'd look back, sensing us I think, but we stayed hidden amid the bushes and trees. His eyesight was poor.

Now and then he stopped to search out blackberries and the red berries that had not yet ripened. His face scrunched up like an owl's when he tasted their tart juices. Like a little bear, he moved on then, singing to himself, taking all the time in the world, plucking the berries gingerly to avoid scratching his plump hands and wrists in the awful tangle of daggers and claws in which the berries rested. Sometimes his hand and arm got caught on the curved hooks of the thorns, and he'd be stuck as though in a trap. He'd wince as he pulled his hand free of the daggers, and as he pulled, other thorns would catch him more firmly; he'd pull harder. Once free, he sucked the blood from his pinprick wounds.

And when he'd had his fill of berries and was nearing the end of the road, he began to pick blossoms, stuffing them like coins into the pockets of his shirt and the baggy shorts he always wore — camper-style shorts with zip-up compartments and all sorts of rings and hooks for hanging compasses and flashlights.

Then he walked down to the big pond we called Hidden Lake, deep in the woods, and sprinkled the white blossoms onto the surface of the muddy water. Frogs would cry out in alarm, leaping from shore's edge with frightened chirps. A breeze would catch the floating petals and carry them across the lake like tiny boats. Swamp Boy would walk up and down the shore, trying to catch those leaping frogs.

Leopard frogs: *Rana utricularia.*

We followed him like jackals, like soul scavengers. We made the charge about once a week: we'd shout and whoop and chase him down like lions on a gazelle, pull that sweet boy down and truss him up with rope and hoist him into a tree. I never touched him. I always held back, only pretending to be in on it. I thought that if I touched him, he would burn my fingers. We knew he was alien, and it terrified us.

With our hearts full of hate, a terrible, frantic, weak, rotting-through-the-planks hate, we — they, the other boys — would leave him hanging there, red-faced and congested, thick-tongued with his upside-down blood, until the sheer wet weight of the sack of blood that was his body allowed him to slip free of the ropes and fall to the ground like a dead animal, like something dripped from a wet limb. But before his weight let him fall free, the remaining flowers he'd gathered fluttered from his pockets like snow.

Some of the boys would pick up a rock or a branch and throw it at him as we retreated, and I was sickened by both the sound of the thuds as the rocks struck his thick body and by the hoots of pleasure, the howls of the boys whenever one of their throws found its target. Once they split his skull open, which instantly drenched his hair, and we ran like fiends, believing we had killed him. But he lived, fumbling free of the ropes to make his way home, bloody-faced and red-crusted. Two days after he got stitched up at the hospital, he was back in the woods again, picking berries and blossoms, and even the dumbest of us could see that something within him was getting strong, and that something in us was being torn down.

Berry blossoms lined the road along which we walked each afternoon — clumps and piles of flowers, each mound of them indicating where we'd strung him up earlier. I started to feel bad about what I was doing, even though I was never an attacker. I merely ran with the other boys for the spectacle, to observe the dreamy phenomenon of Swamp Boy.

One night in bed I woke up with a pain in my ribs, as if the rocks had been striking me rather than him, and my mouth tasted of berries, and I was frightened. There was a salty, stinging feeling of thorn scratches across the backs of my hands and forearms. I have neglected to say that we all wore masks as we stalked and chased him, so he was probably never quite sure who we were — he with those thick glasses.

I lay in the darkness and imagined that in my fright my heart would begin beating faster, wildly, but instead it slowed down. I waited what seemed like a full minute for the next beat. It was stronger than my heart had ever beaten before. Not faster, just stronger. It kicked once, as if turning over on itself. The one beat — I could feel this distinctly — sent the pulse of blood all through my body, to the ends of my hands and feet. Then, after what seemed again like a full minute, another beat, one more round of blood, just as strong or stronger. It was as if I had stopped living and breathing, and it was the beat of the earth's heart in my hurt chest. I lay very still, as if pinned to the bed by a magnet.

The next day we only spied on him, and I was glad for that. But even so, I awakened during the night with that pain in my chest again. This time, though, I was able to roll free of the bed. I went to the kitchen for a glass of water, which burned all the way down as I drank it.

I got dressed and went out into the night. Stars shone through the trees as I walked toward the woods that lay between our subdivision and the school, the woods through which Swamp Boy passed each day. Rabbits sat hunched on people's front lawns like concrete ornaments, motionless in the starlight, their eyes glistening. The rabbits seemed convinced that I presented no danger to them, that I was neither owl nor cat. The lawns were wet with dew. Crickets called with a kind of madness, or a kind of peace.

I headed for the woods where we had been so cruel, along the lazy curves of the bayou. The names of the streets in our subdivision were Pine Forest, Cedar Creek, Bayou Glen, and Shady River, and for once, with regard to that kind of thing, the names were accurate. I work in advertising now, at the top of a steel-and-glass skyscraper from which I stare out at the flat gulf coast, listening to the rain, when it comes, slash and beat against the office windows. When the rain gives way to sun, I'm so high up that I can see to the curve of the earth and beyond. When the sun burns the steam off the skin of the earth, it looks as if the whole city is smoldering.

Those woods are long gone now, buried by so many tons of houses and roads and other sheer masses of concrete that what happened there when I was a child might as well have occurred four or five centuries ago, might just as well have been played out by Vikings in horn helmets or red natives in loincloths.

There was a broad band of tallgrass prairie — waist high, bending gently — that I would have to cross to get to the woods. I had seen deer leap from their beds there and sprint away. I could smell the faraway, slightly sweet odor of a skunk that had perhaps been caught by an owl at the edge of the meadow, for there were so many skunks in the meadow, and so many owls back in the woods. I moved across the silver field of grass in starlight and moonlight, like a ship moving across the sea, a small ship with no others out, only night.

Between prairie and woods was a circle of giant ancient oaks. You could feel magic in this spot, could feel it rise from centuries below and brush against your face like the cool air from the bottom of a deep well. This "buffalo ring" was the only evidence that a herd of buffalo had once been held at bay by wolves, as the wolves tried, with snarling feints and lunges, to cut one of the members out of the herd. The buffalo had gathered in a tight circle to make their stand, heads all facing outward; the weaker ones had taken refuge in the center. Over and over, the sun set and the moon rose and set, again and again, as the wolves kept them in this standoff. Heads bowed, horns gleaming, the buffalo trampled the prairie with their hooves, troughed it up with nose-stinging nitrogen piss and shit in their anger and agitation.

Whether the wolves gave up and left, or whether they darted in, grabbed a leg, and pulled out one of the buffalo — no matter, for all that was important to the prairie, there at the edge of the woods along the slow bayou, was what had been left behind. Over the years, squirrels and other animals had carried acorns to this place, burying them in the rich circular heap of shit-compost. The trees, before they were cut down, told this story.

Swamp Boy could feel these things as he moved across the prairie and through the woods, there at the edge of that throbbing, expanding city, Houston. And I could too, as I held my ribs with both arms because of the strange soreness. I began going into the woods every night, as if summoned.

I would walk the road he walked. I would pass beneath

the same trees from which we had hung him, the limbs thick and branching over the road: the hanging trees. I would walk past those piles of flower petals and berry blossoms, and shuffle my feet through the dry brown oak leaves. Copperheads slept beneath the leaves, cold and sluggish in the night, and five-lined skinks rattled through brush piles, a sound like pattering rain. Raccoons loped down the road ahead of me, looking back over their shoulders, their black masks smudged across their delicate faces.

I would walk past his lake. The shouting frogs fled at my approach. The water swirled and wriggled with hundreds of thousands of tadpoles — half-formed things that were neither fish nor frog, not yet of this world. As they swirled and wriggled in the moonlight, it looked as if the water were boiling.

I would go past the lake, would follow the thin clay road through the starlit forest to its end, to a bluff high above the bayou, the round side of one of the meandering S-cuts that the bayou had carved. I know some things about the woods, even though I live in the city, have never left this city. I know some things that I learned as a child just by watching and listening — and I could use those things in my advertising, but I don't. They are my secrets. I don't give them away.

I would stand and watch and listen to the bayou as it rolled past, its gentle, lazy current always murmuring, always twenty years behind. Stories from twenty years ago, stories that had happened upstream, were only just now reaching this spot.

Sometimes I feel as if I've become so entombed that I have *become* the giant building in which I work — that it is my shell, my exoskeleton, like the seashell in which a fiddler crab lives, hauling the stiff burden of it around for the rest of his days. The chitin of things not said, things not done.

I would stand there and hold my hurt ribs, feel the breezes, and look down at the chocolate waters, the star glitter reflecting on the bayou's ripples, and I would feel myself fill slowly and surely with a strength, a giddiness that urged me to *jump, jump, jump.* But I would hold back, and instead would watch the bayou go drifting past, carrying its story twenty more years down the line, and then thirty, heading for the gulf, for the shining waters.

Then I'd walk back home, undress, and crawl back into bed and sleep hard until the thunder rattle of the alarm clock woke me and my parents and my brothers began to move about the house. I'd get up and begin my new day, the real day, and my ribs would be fine.

I had a secret. My heart was wild and did not belong among people.

I did what I could to accommodate this discrepancy.

We continued to follow him, through the woods and beyond. Sometimes we would spy on him at his house early in the evenings. We watched him and his family at the dinner table, watched them say grace, say amen, then eat and talk. It wasn't as if we were homeless or anything — this was back when we all still had both our parents, when almost everyone did — but still, his house was different. The whole house itself seemed to come alive when the family was inside it, seemed to throb with a kind of strength. Were they taking it from us as we watched them? Where did it come from? You could feel it, like the sun's force.

After dinner they gathered around the blue light of the

television. Our spying had revealed to us that *Daniel Boone* was Swamp Boy's favorite show. He wore a coonskin cap while he watched it, and his favorite part was the beginning, when Dan'l would throw the tomahawk and split the tree trunk as the credits rolled. This excited him so much each time that he gave a small shout and jumped in the air.

After *Daniel Boone,* it would be time for Swamp Boy's mother and father to repair his glasses, if we'd broken them that day. They'd set the glasses down on a big long desk and glue them, or put screws in them, using all sorts of tape and epoxy sealers, adjusting and readjusting them. Evidently his parents had ordered extra pairs, because we had broken them so often. Swamp Boy stood by patiently while his parents bent and wiggled the earpieces to fit him.

How his parents must have dreaded the approach of three in the afternoon, wondering, as it drew near the time for him to get out of school and begin his woods walk home, whether today would be the day that the cruel boys would attack their son. What joy they felt when he arrived home unscathed, back in the safety of his family.

We grew lean through the spring as we chased him toward the freedom of summer. I was convinced that he was absorbing all of our strength with his goodness, his sweetness. I could barely stand to watch the petals spill from his pockets as we twirled him from the higher and higher limbs, could barely make my legs move as we thundered along behind him, chasing him through the woods.

I avoided getting too close, would not become his friend, for then the other boys would treat me as they treated him.

But I wanted to watch.

In May, when Hidden Lake began to warm up, Swamp Boy would sometimes stop off there to catch things. The water was shallow, only neck deep at the center, full of gars, snakes, fish, turtles, and rich bayou mud at its bottom. It's gone now. The trees finally edged in and spread their roots into that fertile swamp bottom, taking it quickly, and no sooner had the trees claimed the lake than they were in turn leveled to make way for what came next — roads, a subdivision, making ghosts of the forest and the lake.

Swamp Boy kept a vegetable strainer and an empty jar in his lunch box. He set his tape-mended glasses down on a rotting log before opening his lunch box, flipping the clasps on it expertly, like a businessman opening his brief-case. He removed his shoes and socks and wiggled his feet in the mud. When his glasses were off like that, we could creep to within twenty or thirty feet of him.

A ripple blew across the water — a slight mystery in the wind or a subtle swamp movement just beneath the surface. I could feel some essence, a truth, down in the soil beneath my feet — but I'd catch myself before saying to the other boys, "Let's go." Instead of jumping into the water or giving myself up to the search for whatever that living essence was beneath me, I watched.

He crouched down, concentrating, looking out over the lake and those places where the breeze had made a little ring or ripple. Then came the part we were there to see, the part that stunned us: Swamp Boy's great race into the water. Building up a good head of steam, running fast and flat-footed in his bare feet, he charged in and slammed his vegetable strainer down into the reeds and rushes. Just as quickly he was back out, splashing, stumbling, having scooped up a big red wad of mud. He emptied the con-tents onto the ground. The mud wriggled with life, all the creatures writhing and gasping, terrible creatures with bony spines or webbed feet or pincers and whiskers.

After carefully sorting through the tadpoles — in various stages of development; half frog and half fish, looking human almost, like little round-headed human babies — angry catfish, gasping snapping turtles, leaping newts, and hellbenders, he put the catfish, the tadpoles, and a few other grotesqueries in his jar filled with swamp water, and then picked up all the other wriggling things and threw them back into the lake.

Then he wiped his muddy feet off as best as he could, put his shoes and socks and strainer in his lunch box, and walked the rest of the way home barefoot. From time to time he held the jar up to the sun, to look at his prizes swimming around in that dirty water. The mud around his ankles dried to an elephant-gray cake. We followed him to his house at a distance, as if escorting him.

That incredible force field, a wall of strength, when he disappeared into his house, into the utility room to wash up — the whole house glowed with it, something emanated from it. And once again I could feel things, lives and stories, meaningful things, stirring in the soil beneath my feet.

I continued to walk out to the woods each night, awakening with a pain so severe in my chest and ribs, a pain and a hunger both, that I could barely breathe.

Could I run out and catch a frog or a tadpole, launch myself wild-assed into the muddy water? Could I bring a shovel out to the prairie one night and dig down, deep down, in search of an old buffalo skull that would still smell rank and earthy but gleam white in the moonlight when I pulled it up? If I had intended to do any of those things, if I had dared to — if I had had the strength and the courage — I should have done so then.

In the evenings, after spying on him as he watched TV, we'd go home to our own suppers, then return and look in on him again. We wouldn't devil him, just watch him. We'd line up, a couple of us at each window, and peer in from the darkness like raccoons.

In his room Swamp Boy had six aquariums set beneath neon lights. He kept the other lights turned off, so that all you could see were his catfish and the hellish tadpoles. There were filters and air pumps bubbling away in those aquariums, humming softly. The water was so clear that it must have seemed like heaven to those poor rescued creatures that had been living in a Houston mud hole.

The catfish were pretty, as were even the feather-gilled tadpoles — morpho-frogs whose hind legs trailed uselessly behind them. He went from tank to tank, bending over to examine the creatures with his patched-up glasses that made him look like a little surgeon. He pressed his nose against the glass and stared in wonder, open-mouthed, touching the sides of the aquarium with his fingers while the sleek, wild-whiskered catfish and bulge-bellied tadpoles circled and swarmed in lazy schools, rising and falling as if with purpose. He tried to count his charges. We'd see him point at them with his index finger, saying the words out loud or to himself, softly, "One, two, three, four . . ."

There was a bottle of aspirin on his desk and a heater in each of the tanks, for cold winter nights. Whenever he suspected that a catfish or a tadpole was feeling ill, he'd drop an aspirin in the water. It would make a cloudy trail as it fell.

Sometimes he'd lie in bed with his hands behind his head and watch the fish and tadpoles go around and around in their new home. When the moon was up and the lights were off it looked as though his room were under water — as if *he* were under water among the catfish.

God, we were devils. It occurred to a couple of the older boys to see how far he could run. Usually we caught him and strung him up fairly quickly, after only a short chase, but one day we tried to run him to exhaustion, to try and pop his fat, strong heart.

After school we put on wolf masks and made spiked collars by driving nails through leather dog collars, which we fastened around our necks. We spoke to one another in snarling laughs, our voices muffled through the wolf masks.

We started out after him the minute he hit the woods, bending our heads low to the ground and pretending to sniff his scent, howling, trotting along behind him, loping and barking. We chased him through the woods and down along the bayou on the other side of a forested ridge. In his fear he started making sounds like a lost calf. There was cane along the bayou, flood-killed, dry-standing ghost bamboo, and Swamp Boy plowed through that as if going through a dead corn field, snapping the bamboo in all directions, running as if the forces of hell had opened up.

All we were going to do was throw mud on him, once he got too tired to go any farther. Roll him around in the mud a little, maybe. And break his glasses, of course.

But this time he was really afraid.

It was exciting, chasing him through the tiger-slash stripes of light, following the swath of his flight through all that knocked-over dead bamboo. It was about the most exciting thing we had ever done.

We chased him to the small bluff overlooking the bayou, and Swamp Boy paused for only the briefest of seconds before making a Tarzan dive into the milky brown water. He swam immediately for the slick clay bank on the other side, toward north Houston where the rich people lived, where I imagined he would skulk up to some rich person's back yard, shivering, shoeless, smelling like some vile swamp thing, waiting until dark, so large was his shame. He'd hide in their bushes, perhaps, before creeping up to the back porch — still dripping wet and muddy, and bloody from where the canes had stabbed him — and then, crying, ask if he could use the telephone.

If this were not all a lie, a re-creation or manipulation of the facts, and if I were the boy who had chased the other boy through the cane, rather than the boy who had leapt into the muddy bayou, then what I would have done, what I should have done, was something heroic: I would have held out my hands like an Indian chief, stopping the other boys from jumping in and swimming after him, or even from gloating. I would have said something noble, like, "He got away. Let him go."

I might even have gone home and called Swamp Boy's parents, so that he wouldn't have to lurk in the shadows in some rich person's yard — afraid to walk home through the woods because of the masked gang, but also afraid to go ask to use the phone.

That's what I'd have done if I were the boy who chased him, rather than the boy who got chased, and who made that swim. Who kept, and worshiped, those baroque creatures in his aquarium.

I was that boy, and I was the other one too. I was at the edge of fear, the edge of hesitancy, and had not yet — not then — turned back from it.

There's a heavy rain falling today. The swamps are writhing with life.

Fires

SOME YEARS the heat comes in April. There is always wind in April, but with luck there is warmth too. When the wind is from the south, the fields turn dry and everyone in the valley moves his seedlings outdoors. Root crops are what do best up here. The soil is rich from all the many fires, and potatoes from this valley taste like candy. Carrots pull free of the dark earth and taste like crisp sun. Strawberries do well here if they're kept watered.

The snow has left the valley by April, has moved up into the surrounding woods, and then by July the snow is above the woods, retreating to the cooler, shadier places in the mountains. But small oval patches of it remain behind. As the snow moves up into the mountains, snowshoe hares, gaunt but still white, descend on the garden's fresh berry plants. You can see the rabbits, as white as Persian cats, from a mile away, coming after your plants, hopping through sun-filled woods and over rotting logs, following centuries-old game trails of black earth.

The rabbits come straight for my outside garden like relentless zombies, and I sit on the back porch and sight in on them. But they are too beautiful to kill in great numbers. I shoot only one every month or so, just to warn them. I clean the one I shoot and fry it in a skillet with onions and half a piece of bacon.

At night when I'm restless, I go from my bed to the window and look out. In spring I see the rabbits standing at attention around the greenhouse, aching to get inside. Several of them will dig at the earth, trying to tunnel in, while others sit there waiting.

Once the snow is gone, the rabbits begin to lose their white fur — or rather, they do not lose it, but it begins to turn the mottled brown of decaying leaves. Finally the hares are completely brown, and safe again, indistinguishable from the world around them.

I haven't lived with a woman for a long time. Whenever one does move in with me, it feels as if I've tricked her, caught her in a trap, as if the gate has been closed behind her, and she doesn't yet realize it. It's very remote up here.

One April a runner came to the valley to train at altitude. She was the sister of my friend Tom. Her name was Glenda, and she was from Washington State. Glenda had run races in Italy, France, and Switzerland. She told everyone, including the rough loggers and their wives, that this was the most beautiful place she had ever seen, and we believed her. Very few of us had ever been anywhere else to be able to question her.

We often sat at the picnic tables in front of the saloon, ten or twelve of us at a time, half of the town, and watched the river. Ducks and geese, heading north, stopped in our valley to breed, build nests, and raise their young. Ravens, with their wings and backs shining greasy in the sun, were always flying across the valley, from one side of the mountains to the other. Anyone who needed to make a little money could always do so in April by planting seedlings for the Forest Service, and it was an easier time because of that fact, a time of no bad tempers, of worries put aside for a while. I did not need much money, in April or in any other month, and I would sit at the picnic table with Glenda and Tom and Nancy, Tom's wife, and drink beer. Glenda had yellow hair that was cut short, and lake-blue eyes, a pale face, and a big grin, not unlike Tom's, that belied her seriousness, though now that she is gone, I remember her always being able to grin *because* of her seriousness. Like the rest of us, Glenda had no worries, not in April and certainly not later on, in the summer. She had only to run. She was separated from her boyfriend, who lived in California, and she didn't seem to miss him, didn't ever seem to think about him.

Before planting the seedlings, the Forest Service burned the slopes they had cut the previous summer and fall. In the afternoons there would be a sweet-smelling haze that started about halfway up the valley walls, then rose into the mountains and spilled over them, moving north into Canada, riding on the south winds. The fires' haze never settled in our valley, but would hang just above us, turning the sunlight a smoky blue and making things, when seen across the valley — a barn in another pasture or a fence line — seem much farther away than they really were. It made things seem softer, too.

Glenda had a long scar on the inside of her leg that ran from the ankle all the way up to mid-thigh. She had injured her knee when she was seventeen. This was before arthroscopic surgery, and she'd had to have the knee rebuilt the old-fashioned way, with blades and scissors, but the scar only seemed to make her legs, both of them, look even more beautiful. The scar had a graceful curve to it as it ran the distance up her leg.

Glenda wore green nylon shorts and a small white T-shirt when she ran, and a headband. Her running shoes were dirty white, the color of the road dust during the dry season.

"I'm thirty-two and have six or seven more good years of running," she said whenever anyone asked her what her plans were, why she ran so much, and why she had come to our valley to train. Mostly it was the men who asked, the ones who sat with us in front of the saloon watching the river, watching the spring winds move across the water. We were all glad that winter was over. Except for Nancy, I do not think the women liked Glenda very much.

It was not well known in the valley what a great runner Glenda was. And I think it gave Glenda pleasure that it wasn't.

"I'd like you to follow Glenda on a bicycle," Tom said the first time I met her. He'd invited me over for dinner a short time after she'd arrived. "There's money available from her sponsor to pay you for it," he said, handing me some money, or trying to, finally putting it in my shirt pocket. He had been drinking, and seemed as happy as I had seen him in a long time. After stuffing the bills into my pocket, he put one arm around Nancy, who looked embarrassed for me, and the other arm around Glenda, who did not, and so I had to keep the money, which was not that much anyway.

"You just ride along behind her with a pistol." Tom had a pistol holstered on his belt, a big one, and he took it off and handed it to me. "And you make sure nothing happens to her, the way it did to that Ocherson woman."

The woman named Ocherson had been walking home

along the river road after visiting friends when a bear evidently charged out of the willows and dragged her across the river. She had disappeared the previous spring, and at first everyone thought she had run away. Her husband had gone around all summer making a fool of himself, bad-mouthing her. Then hunters found her body in the fall, right before the first snow. Every valley had its bear stories, but we thought our story was the worst, because the victim had been a woman.

"It'll be good exercise for me," I said to Tom, and then I said to Glenda, "Do you run fast?"

It wasn't a bad job. I was able to keep up with her most of the time. Some days Glenda ran only a few miles, very fast, and on other days it seemed that she ran forever. There was hardly ever any traffic — not a single car or truck — and I daydreamed as I rode along behind her.

Early in the morning we'd leave the meadow in front of Tom's place and head up the mountain on the South Fork road, above the river and into the woods, going past my cabin. Near the summit, the sun would be up and burning through the haze of the planting fires. Everything would look foggy and old, as if we had gone back in time and not everything had been decided yet.

By the time we reached the summit, Glenda's shirt and shorts were drenched, her hair damp and sticking to the sides of her face; her socks and even her running shoes were wet. But she always said that the people she would be racing against would be training even harder than she was.

There were lakes around the summit, and the air was cooler. On the north slope the lakes still had a thin crust of ice over them, a crust that thawed each afternoon but froze again at night. What Glenda liked to do after she'd reached the summit — her face flushed and her wrists limp and loose, so great was the heat and her exhaustion — was to leave the road and run down the game trail, tripping, stumbling, running downhill again. I would have to throw the bike down and hurry after her. She'd pull her shirt off and run into the shallows of the first lake she saw, her feet breaking through the thin ice. Then she'd sit down in the cold water like an animal chased there by hounds.

"This feels good!" she said the first time she did that. She leaned her head back on the shelf of ice behind her and spread her arms as if she were resting on a cross. She looked up through the haze at the empty sky above the tree line.

"Come over here," she said. "Come feel this."

I waded out, following her trail through the ice, and sat down next to her.

She took my hand and put it on her chest.

What I felt was like nothing I had ever imagined. It was like lifting up the hood of a car with the engine on and seeing all the cables and belts and fan blades still running. Right away, I wanted to get her to a doctor. I wondered, if she were going to die, whether I would be held responsible. I wanted to pull my hand away, but she made me keep it there, and gradually the drumming slowed, became steadier, and still she made me keep my hand there, until we could both feel the water's coldness. Then we got out. I had to help her up because her injured knee was stiff. We spread out our clothes and lay down on flat rocks to dry in the wind and the sun. She'd said that she had come to the mountains to run because it would strengthen her knee. But there was something that made me believe that that was not the truth, though I cannot tell you what other reason there might have been.

On every hot day we went into the lake after her run. It felt wonderful, and lying in the sun afterward was wonderful too. Once we were dry, our hair smelled like the smoke from the planting fires. There were times when I thought that Glenda might be dying, and had come here to live out her last days, to run in a country of great beauty.

By the time we started our journey home, there'd be a slow wind coming off the river. The wind cleared a path through the haze, moving it to either side, and beneath it, in that space between, we could see the valley, green and soft. Midway up the north slope, the ragged fires would still be burning. Wavering smoke rose from behind the trees.

The temptation to get on the bike and coast all the way down was always strong, but I knew what my job was, we both did. It was the time when bears came out of hibernation, and the safety of winter was not to be confused with the seriousness of summer, with the way things were changing.

Walking back, we would come upon ruffed grouse, the males courting and fanning in the middle of the road, spinning in a dance, their throat sacs inflated and pulsing bright red. The grouse did not want to let us go past: they stamped their feet and blocked our way, trying to protect some small certain area they had staked out for themselves. Glenda stiffened whenever she saw the fanning males, and shrieked when they rushed in and tried to peck at her ankles.

We'd stop at my cabin for lunch, and I'd open all the windows. By then the sun would have heated the log walls, and inside was a rich dry smell, as there is when you have been away from your house for a long time. We would sit at the breakfast room table and look out the window at the weedy chicken house I'd never used and at the woods going up the mountain behind the chicken house. We drank coffee and ate whitefish, which I had caught and smoked the previous winter.

I had planted a few young apple trees in the back yard that spring, and the nursery that had sold them to me said that these trees could withstand even the coldest winters, though I wasn't sure I believed it. They were small trees and would not bear fruit for four years, and that had sounded to me like such a long time that I really had to think about it before buying them. But I bought them anyway, without really knowing why. I also didn't know what would make a person run as much as Glenda did. I liked riding with her, though, and having coffee with her after the runs, and I knew I would be sad to see her leave the valley. I think that was what kept up the distance between us, a nice distance — the fact that both of us knew that she would stay only a short time, until the end of August. Knowing this seemed to take away any danger, any wildness. It was a certainty; there was a wonderful sense of control.

I had a couple of dogs in the back yard, Texas hounds that I'd brought up north with me a few years before. I kept them penned up in the winter so they wouldn't chase deer, but in the spring and summer I let them lie around in the grass, dozing. There was one thing they would chase, though, in the summer. It lived under the chicken house, and I don't know what it was; it ran too fast for me to ever get a good look at it. It was small and dark with fur, but it wasn't a bear cub. Perhaps it was some rare animal, something from Canada — maybe something no one had ever seen before. Whatever it was, it never grew from year to year, yet it seemed young somehow, as if it might someday grow. It would rip out of the woods, a fleet blur headed for its burrow, and as soon as the dogs saw it, they would be

up and baying, right on its tail, but the thing always reached
its burrow under the chicken house just ahead of them.

Glenda and I would sit at the window and watch for
it every day. But it kept no timetable, and there was no
telling when it would come, or even if it would. We called
it a hedgehog, because that was the closest thing it might
have resembled.

Some nights Glenda would call me on the shortwave
radio. She would key the mike a few times to make it crackle
and wake me up, and then I would hear a mysterious voice
floating in static through my cabin. "Have you seen the
hedgehog?" she would ask sleepily, but it was never her
real voice there in the dark with me. "Did you see the
hedgehog?" she'd want to know, and I'd wish she were with
me at that moment. But it would be no good; Glenda was
leaving in August, or September at the latest.

"No," I'd say in the dark. "No hedgehog today. Maybe
it's gone away." Though I had thought that many times,
I would always see it again, just when I thought I never
would.

"How are the dogs?" she'd ask.

"They're asleep."

"Good night," she'd say.

"Good night."

One Thursday night I had Tom and Nancy and Glenda
over for dinner. Friday was Glenda's day off from running,
so she allowed herself to drink and stay up late on Thurs-
days. Before dinner, we started out drinking at the saloon.
Around dusk we went down to my cabin, and Glenda and
I fixed dinner while Tom and Nancy sat on the front porch,
watching the elk appear in the meadow across the road as
the light faded.

"Where's this famous hedgehog?" Tom bellowed, puffing a cigar, blowing smoke rings into the night, big perfect O's. The elk lifted their heads, chewing the summer grass like cattle, the bulls' antlers glowing with velvet.

"In the back yard," Glenda said as she washed the salad greens. "But you can only see him in the daytime."

"Aww, bullshit!" Tom roared, standing up with his bottle of Jack Daniel's. He took off down the steps, stumbling, and the three of us put down what we were doing to get flashlights and run after him to make sure he was all right. Tom was a trapper, and it riled him to think there was an animal he did not know, could not trap, could not even see. Out by the chicken house, he got down on his hands and knees, breathing hard, and we crowded around him to shine the flashlights into the deep, dusty hole. He made grunting noises that were designed, I suppose, to make the animal want to come out, but we never saw anything. It was cold under the stars. Far off, the planting fires burned, but they were held in check, controlled by back-fires.

I had a propane fish fryer, and we put it on the front porch and cut the trout into cubes, rolled them around in flour, then dropped them in hot, spattering grease. We fixed about a hundred trout cubes, and when we finished eating there were none left. Glenda had a tremendous appetite, and ate almost as many as Tom. She licked her fingers afterward, and asked if there were any more.

After dinner we took our drinks and sat on the steep roof of my cabin, above the second-floor dormer. Tom sat out on the end of the dormer as if it were a saddle, and Glenda sat next to me for warmth, and we watched the fires spread across the mountainside, raging but contained. Below us, in the back yard, those few rabbits that still had not turned completely brown began coming out of the woods. Dozens of them approached the greenhouse, then stopped and lined up around it, wanting to get into the tender young carrots and the Simpson lettuce. I had put sheets down on the ground to trick them, and we laughed as the rabbits shifted nervously from sheet to sheet, several of them huddling together on one sheet at a time, imagining they were protected.

"Turn back, you bastards!" Tom shouted happily. That woke the ducks on the pond nearby, and they began clucking among themselves. It was a reassuring sound. Nancy made Tom tie a rope around his waist and tie the other end around the chimney, in case he fell. But Tom said he wasn't afraid of anything, and was going to live forever.

Glenda weighed herself before and after each run. I had to remind myself not to get too close to her; I only wanted to be her friend. We ran and rode in silence. We never saw any bears. But she was frightened of them, even as the summer went on without us seeing any, and so I always carried the pistol. We had gotten tanned from lying out by the lake up at the summit. Glenda took long naps at my cabin after her runs; we both did, Glenda sleeping on my couch. I'd cover her with a blanket and lie down on the floor next to her. The sun would pour in through the window. There was no longer any other world beyond our valley — only here, only now. But still, I could feel my heart pounding.

It turned drier than ever in August, and the loggers began cutting again. The days were windy, and the fields and meadows turned to crisp hay. Everyone was terrified of sparks, especially the old people, because they'd seen big fires rush through the valley, moving through like an army: the big fire in 1901, and then the monstrous one in 1921 that burned up every tree except for the luckiest ones, so that for years afterward the entire valley was barren and scorched.

One afternoon in early August Glenda and I went to the saloon. She lay down on top of a picnic table and looked up at the clouds. She would be going back to Washington in three weeks, she said, and then down to California. Almost all of the men would be off logging in the woods by then, and we would have the whole valley to ourselves. Tom and Nancy had been calling us "the lovebirds" since July, hoping for something to happen — something other than what was, or wasn't, happening — but they'd stopped in August. Glenda was running harder than ever, really improving, so that I was having trouble keeping up with her.

There was no ice left anywhere, no snow, not even in the darkest, coolest parts of the forest, but the lakes and rivers were still ice cold when we waded into them. Glenda continued to press my hand to her breast until I could feel her heart calming, and then almost stopping, as the waters worked on her.

"Don't you ever leave this place," she said as she watched the clouds. "You've got it really good here."

I stroked her knee with my fingers, running them along the inside scar. The wind tossed her hair around. She closed her eyes, and though it was hot, there were goose bumps on her tanned legs and arms.

"No, I wouldn't do that," I said.

I thought about her heart, hammering in her chest after those long runs. At the top of the summit, I'd wonder how anything could ever be so *alive*.

The afternoon she set fire to the field across the road from my cabin was a still day, windless, and I suppose that Glenda thought it would do no harm — and she was right, though I did not know it then. I was at my window when I saw her out in the field lighting matches, bending down and cupping her hands until a small blaze appeared at her feet. Then she came running across the field.

At first I could hardly believe my eyes. The smoke in front of the fire made it look as if I were seeing something from memory, or something that had happened in another time. The fire seemed to be secondary, even inconsequential. What mattered was that she was running, coming across the field toward my cabin.

I loved to watch her run. I did not know why she had set the fire, and I was very afraid that it might cross the road and burn up my hay barn, even my cabin. But I was not as frightened as I might have been. It was the day before Glenda was going to leave, and mostly I was delighted to see her.

She ran up the steps, pounded on my door, and came inside, breathless, having run a dead sprint all the way. The fire was spreading fast, even without a wind, because the grass was so dry, and red-winged blackbirds were flying out of the grass ahead of it. I could see marsh rabbits and mice scurrying across the road, heading for my yard. It was late in the afternoon, not quite dusk. An elk bounded across the meadow. There was a lot of smoke. Glenda pulled me by the hand, taking me back outside and down the steps, back out toward the fire, toward the pond on the far side of the field. It was a large pond, large enough to protect us, I hoped. We ran hard across the field, and a new wind suddenly picked up, a wind created by the flames. We got to the pond and kicked our shoes off, pulled off our shirts and jeans, and splashed into the water. We waited for the flames to reach us, and then work their way around us.

It was just a grass fire. But the heat was intense as it rushed toward us, blasting our faces with hot wind.

It was terrifying.

We ducked our heads under the water to cool our drying faces and threw water on each other's shoulders. Birds flew past us, and grasshoppers dived into the pond with us, where hungry trout rose and snapped at them, swallowing them like corn. It was growing dark and there were flames all around us. We could only wait and see whether the grass was going to burn itself up as it swept past.

"Please, love," Glenda was saying, and I did not understand at first that she was speaking to me. "Please."

We had moved out into the deepest part of the pond, chest deep, and kept having to duck beneath the surface because of the heat. Our lips and faces felt scorched. Pieces of ash were floating down to the water like snow. It was not until nightfall that the flames died, leaving just a few orange ones flickering here and there. But the rest of the small field was black and smoldering.

It turned suddenly cold, and we held on to each other tightly, because we were shivering. I thought about luck and about chance. I thought about fears, all the different ones, and the things that could make a person run.

She left at daylight. She would not let me drive her home — she said she wanted to run instead, and she did. Her feet raised puffs of dust in the road.

The Valley

ONE DAY I left the South, fled my job, and ran to the heart of snow, the far Northwest. I live in a cabin with no electricity, and I'm never leaving.

There aren't many people in this valley — twenty-six registered voters — and rather than disliking almost everyone, as I found it so easy to do in the city, I can now take time to love practically everyone.

I have to start small. I have to get it right.

Jody Michaels is sixty years old and lives up in the woods. She takes in stray dogs that come her way. There are more than you'd think: they jump out of the backs of trucks, or run away from home. They strike out for the North.

Jody's is the last cabin they come to before going over into Canada. She keeps a large team of sled dogs — huskies and malamutes, blue-eyed creatures that have so much wolf in them that they don't know how to bark, and can only howl.

When the moon comes up over the mountains that ring

our small bowl of a valley like a high fence, all of Jody's dogs begin to howl, a sound that echoes around the valley. Perhaps the wild, strong sled dogs attract the strays. Often we'll see a stray loping down the road, dragging a broken leash, a broken chain — usually a big dog — and it'll be heading north, for Jody's.

She keeps them in a holding pen for one week, puts a note on the blackboard in front of the Mercantile, and calls the vet in Libby, which is due southwest, sixty miles across the snowdrift-covered pass. If no one claims the dog after that week, she does a very odd thing: she hikes to the top of Hensley Mountain with the dog and sits down with it.

You can see the whole valley from the top of Hensley. It's above the tree line, just barely, and the wind whips and gusts, blows your hair into your eyes. Jody must feel a little like God when she's there surveying things.

And she watches that dog too, watches the way it pants, and the way it looks out at the valley and off toward Canada. Jody knows dogs so well that she can tell, up in that blowing wind, if the dog can survive on its own or not. If it can, she'll unclip the dog's leash, take the collar off, and let it go down the mountain into the deep woods that cross over into Canada — she'll let the dog have its wish. But if Jody doesn't find what she's looking for, she'll lead it back to her cabin. Later in the day she'll drive it to the pound, where, almost always, it will be the end of the line for that runaway dog. There is a man in the valley, never mind his name, who for a dollar will take unwanted dogs up the road a ways and gas them or shoot them. Needless to say, Jody does not employ his services.

I am like those stray dogs, and I think Jody is too. Those dogs have run a long way to get here.

. . .

No one has money in the valley. No one has money even in the little town of Libby. Some of the people who have sled teams rely on road kills to feed their dogs — such large, hungry dogs — and for a fact, you never see a road-killed deer or elk up here. Whenever one of us does strike a deer, instead of leaving it, we load it into the truck (if the truck is still drivable), and head for the Dirty Shame Saloon.

The Dirty Shame sits at the base of Hensley Mountain, which, back in the forties, got a radar dish put on top of it, one of a whole chain of dishes the Air Force had set up along the Northwest's peaks to detect bombers flying over from Russia, the theory being that as we weren't far away — the Russian planes would only have to zip across Alaska, the Yukon, and British Columbia — it would be an easy matter to dive-bomb the valley, riddle the Dirty Shame (which has been here forever) with bullets, and strafe the Mercantile. The radar dish is still there, abandoned, and the lonely dirt road to the top, which seems to lead into the clouds, has long been grown over, crisscrossed with windfall timber and young aspen trees.

One thing from those days did not fall into disrepair, however: the warning siren that was supposed to sound whenever a Russian plane was detected. Handy with tools and electronics, Joe, the owner of the Dirty Shame, decided to hike up there one day and disassemble the siren. He brought it down to the saloon and mounted it on the front porch. Now, every time someone hits a deer and brings it in for barbecuing, Joe shorts the siren's wires with the blade of his pocketknife. Wolves, coyotes, and dogs go crazy when he does that. The siren is so loud that some people in Idaho and Washington can hear it, but because the roads into the valley are in such bad shape, outsiders have no hope of getting here in time for the evening barbecue. Everyone who hears the siren knows what it means.

If it's summertime when the wail goes up, we gather at the saloon around six or six-thirty. Jody comes in her little wagon, pulled by huskies and malamutes. There are nearly as many children as there are registered voters, and after the barbecue the whole group of us will dance until the sunlight leaves, which isn't until around midnight. We bring lettuce from our gardens for the barbecue, and fresh-baked bread. Doug, who is not a veterinarian but is good with animals — he sews them up, and people too, after they've been hurt — brings jars of honey from his beehives. Dave brings his banjo, and Janie her fiddle. Young Terjaney has an enormous electric accordion with row upon row of colored flashing lights which once belonged to his father. Old Mr. Terjaney had brought the accordion all the way from Hungary. He kept it strapped to his chest when he played. The sound was magnificent.

Old Mr. Terjaney drank a lot. Along with everything else, we bring homemade beer to the barbecue, made in our cellars during the slow winters, beer that we keep chilled in the river during summer. Old Mr. Terjaney would open one jar after another of the deep amber–colored brew. He'd get out in the road and dance as he played his big one-footed polkas and waltzes. He kept a jar of beer perched on top of the accordion.

One night, near midnight, he spilled his beer. He'd been dancing and playing a polka with his jar wobbling on top of his accordion. The instrument was hot from the good use he'd made of it, and it exploded like fireworks, electrocuting Old Mr. Terjaney right there in the road. We thought he'd done it on purpose — perhaps this was a special function of the instrument when he pressed a certain button. There were so many buttons. We even cheered at first. It's amazing that Joe was able to repair it.

If it is snowing when you go out to get wood for your fireplace, tie one end of a rope around your waist and tie the other end to the cabin door. The snow can start coming down so fast and hard that in the short time it takes you to get to the woodshed, you can get lost in a whiteout on your way back. It doesn't sound like it's possible, but it happened to me once. A light snow turned heavy in just seconds, and then became a blizzard. I ended up staying in the woodshed all night waiting for daylight. I felt ridiculous, but not as ridiculous as I would have felt dying within a mile of my cabin, when all I had wanted to do was get a few sticks of wood.

There is some compass in all of us that does not want us to walk a straight line. I respect this, and do not try to challenge it in blizzards.

Sometimes people run out of gas (visitors, not locals) up on the pass, where during the winter traffic can go by only every second or third day, and some of them freeze to death in their cars — traveling without heavy clothes, without sleeping bags in the back — and others freeze in the woods when they get out of their cars and try to walk for help. Everyone up here has CB and shortwave radios in their trucks. You can live in a dangerous place quite easily, but to visit it is another thing.

We've got a nice cemetery. There are two cemeteries, actually: one that no one seems to know about, up in the hills above the river, that this kid just found while out walking one day. But the other cemetery, which originally catered mostly to loggers — since they were the ones who used it the most, what with trees falling on them and saws back-bucking and trucks and skidders rolling off cliffs and the like — is now used by everybody, and is majestic.

It's up on Boyd Hill, and you can see the river from it, even through the larch trees, which are centuries old. Two hundred feet tall, they tower like redwoods and have withstood even the biggest fires. They're so huge that eight or ten people holding hands can't encircle them. The larches line all sides of the cemetery's wrought-iron fence, and the air beneath the canopy of trees so high above is a different kind of air, motionless, even when the rest of the woods is windy. Different, too, is the thin light that's able to filter through. Moss grows on the headstones. The shade is cool and smells good. There's a spring nearby, up higher on the mountain.

The timber companies would love to cut the trees around the cemetery — each tree is worth several thousand dollars by itself — but no one starts a chain saw within a mile of the place; it's an unwritten rule.

"Give them a rest for once," says Mack, the little man who takes care of the cemetery, emptying out old flowers, bringing in new ones. It's not the trees he's talking about.

The names of some of the people in the cemetery, if you can believe the headstones, are Piss-Fir Jim, Windy Joe Griff, and Solo Dog Thompson. There was a hermit buried here in the sixties, an old man who even by valley standards was an outsider. He lived as high up in the mountains as you could get, higher than any deer or elk lived, so hunters rarely saw him. Because no one knew his name, he was called The Hermit on his gravestone.

He used to come down twice a year with his mules to buy groceries — flour and beans, mostly. One spring he didn't show up, so Joe, Young Terjaney, and A. C. Rightman went to check on him, and sure enough he'd died during the winter.

It was windy up there, windier than you could imagine, Rightman says. They found the hermit about a mile down the trail leading away from his cabin. The body was frozen in a crouch, as if he'd known he was sick and had been trying to crawl into town for help — though crawling would have taken weeks, and we couldn't imagine what ailment he might have had that would have prevented him from walking but would still let him crawl.

"Probably a bellyache," Rightman will say, stroking his chin, if you ask him. The mules were gone ("Grizzly," Rightman says), but inside the cabin they found the old man's cats, living on mice and melting snow that had dripped in through cracks in the roof during warm May afternoons. All the cats ran out the cabin door except for a large, placid orange one, which Rightman took home to his wife, Marva. Cats can live to a ripe old age in the valley, and dogs can reach the age of twenty or twenty-five; it's not uncommon. Rightman and Marva never gave the cat a proper name. "Hey, hermit's cat," he'll say, "get down off that table." They had the cat for a long time.

For a headstone the hermit has a rough piece of granite pulled off a talus slope, just like the others. John Skabellund, the blacksmith, chiseled the old man's name on it — *The Hermit* — but he's buried off in a corner of the cemetery, as far away as possible from everyone else. It was a joke at first, but I can tell now that a few people feel bad about it.

"You can bury us next to him," Janie says, speaking for her whole family.

The other cemetery is way over on Yellow Creek, up in the mountains. No roads lead to it. Deep woods, grizzlies, and elk surround it, and nothing else. Only women are buried there. The place is a mystery to everyone.

Not many people know how to find it. Rightman says it's run by outsiders, and he must be right. He took me up there on skis one winter. Somehow it feels safer in winter.

The women's pictures — taken when they were young, and framed in glass — are inset in the marble headstones. We ask, how do you carry such heavy stones into the deep woods? The headstones are inscribed with the women's names, their dates of birth and death, and that's all. None of them smiled for their pictures.

Some of the pictures are fading — there are headstones from the 1910s and 1920s. But there are new ones too: the newest is marked just two years ago.

"It's weird," Rightman says. He takes a long drag on his cigarette, finishing it, and flicks it toward the nearest white stone. "It's got to be easterners of some kind." He's probably right. They all look fashionable, like women from New York or Philadelphia. None of them are from out here, we can tell.

Whenever a car drives into the valley — almost always lost — and stops at the Mercantile for gas and directions, those of us who are around will pretend to be interested in helping. But if that car's got eastern license plates, what we're really checking for is where they've got the body hidden.

"Packhorses," Rightman says. "If they die in the winter, they freeze 'em, then bring 'em out here in the spring, at night, and take 'em up there on packhorses."

But it's muddy in the spring, and we never see hoofprints.

Rightman shrugs, draws on his cigarette. He knows he's right. "Got to be packhorses," he says. "That's all there is to it."

My theory is that the women were taken there during a thunderstorm, so the rain would wash away the tracks of those who buried them. I'd like to hide in a cool bower, to see them do it. I want to watch them digging in the rain, the rain beating on their backs.

The other night I saw a cougar run across the road in front of me, flashing across the sweep of my headlights, chasing something.

This is a beautiful valley. I wake up smiling sometimes because I have all my days left to live in this place. I hike up into the hills, to a rock back in the trees, and sit there and just look. On the road far below, a friend drives past in his truck, moving so slowly it seems that a man on foot could walk alongside and still keep up. I watch until the truck disappears around the bend. When dusk comes, purple light slides in from all directions.

The lights in my friends' cabins begin to come on then, glowing patches in the dark.

Antlers

HALLOWEEN brings us closer. The Halloween party at the saloon is when we — all three dozen or so of us — recollect again why we live in this cold, blue valley. Sometimes tourists come when the summer grass is high, and the valley opens up a little. People slip in and out of it; in summer it's almost a regular place. But in October the less hardy of heart leave as the snow begins to fall. It becomes our valley again, and there's a feeling like a sigh, a sigh after the great full meal of summer.

We don't bother with masks at the party because we know one another so well, if not through direct contact, then through word of mouth — what Dick said Becky said about Don, and so on. Instead, we strap horns on our heads, moose antlers, deer antlers, or even the high throwback of elk antlers. We have a big potluck supper and get drunk as hell, even those of us who do not generally drink. We put the tables and chairs outside in the gust-driven snow and put nickels in the juke box and dance until early morning to Elvis, The Doors, or Marty Robbins. Mock battles occur when the men and women bang their antlers against each other. We clomp and sway in the barn.

Around two or three in the morning we drive or ski or snowshoe home, or we ride back on horses — however we got to the party is how we'll return. It usually snows big on Halloween — a foot, a foot and a half. Whoever drove down to the saloon will give the skiers a lift by fastening a long rope to the rear bumper of his truck, and the skiers hold on to that rope, still wearing antlers, too drunk or tired to take them off. Pulled up the hill, gliding silently on the road's hard ice, we keep our heads tucked against the wind and snow. Like children dropped off at a bus stop, we let go of the rope when the truck reaches our dark cabin. It would be nice to be greeted by a glowing lantern in the window, but you don't ever go to sleep or leave with a lantern burning like that — it can burn your cabin down in the night. We return to dark houses, all of us.

The antlers feel natural after having them lashed to our heads for so long. Sometimes we bump them against the door frame going in, and knock them off.

There is a woman up here, Suzie, who has moved through the valley with a rhythm that is all her own. Over the years Suzie has been with all the able-bodied men of the valley. All, that is, except for Randy. He still wishes for his chance, but because he is a bowhunter — he uses a strong compound bow and wicked, heart-gleaming aluminum arrows with a whole spindle of razor blades at one end — she will have nothing to do with him.

At times I wanted to defend him, even though I strongly objected to bowhunting. Bowhunting, it seemed to me, was brutal. But Randy was just Randy, no better or worse than any of the rest of us who had dated Suzie. Wolves eviscerate their prey; it's a hard life. Dead's dead, right? And isn't pain the same everywhere?

Suzie has sandy red hair, high cold cheeks, and fury-blue eyes. She is short, no taller than even a short man's shoulders. Suzie's boyfriends have lasted, on average, for three months. No man has ever left her — even the sworn bachelors among us have enjoyed her company. It is always Suzie who goes away from the men first.

When she settled for me, I'm proud to say that we stayed together for five months, longer than she'd ever been with anyone — long enough for people to talk, and to kid her about it.

Our dates were simple. We'd drive up into the snowy mountains, on those mountains that had roads, as far as we could go before the snow stopped us, and gaze at the valley. Or we'd drive into town, sixty miles away on a one-lane, rutted, cliff-hanging road, for dinner and a movie. I could see how there might not have been enough heat and wild romance in it for some of the other men — there'd been talk — but for me it was warm and right while it lasted.

When she left, I did not think I would ever eat or drink again. It felt as if my heart had been torn from my chest, as if my lungs were on fire; every breath burned. I didn't understand why she had to leave. I'd known it would come someday, but still, it caught me by surprise.

Besides being a bowhunter, Randy is a carpenter. He does odd jobs for people in the valley, usually of the sort that requires fixing up old cabins. He keeps his own schedule, and stops work entirely in the fall so he can hunt to his heart's content. He'll roam the valley for days, exploring the wildest places. We all hunt in the fall — grouse, deer, elk, though we leave the moose and bear alone because they are not as common — but none of us is clever or stealthy enough to bowhunt. With a bow, you have to get close to the animal.

Suzie doesn't approve of hunting in any form. "That's what cattle are for," she said one day in the saloon. "Cattle are like city people. Cattle expect, even deserve, what they've got coming. But wild animals are different. Wild animals enjoy life. They live in the woods on purpose. It's cruel to go in after them and kill them. It's cruel."

We all hoo-rahed her and ordered more beers.

She doesn't get angry, exactly. She understands that everyone hunts here, men and women alike. She knows we love animals, but for one or two months out of the year we also love to hunt them.

Randy is so good at what he does it makes us jealous. He can crawl to within thirty yards of an animal when it is feeding, or he can sit so still that it walks right past him. Once shot, the animal runs but a short way — it bleeds to death or dies from trauma. The blood trail is easy to follow, especially in the snow. No one wants it to happen this way, but there's nothing to be done about it; bowhunting is like that. The others of us look at it as being much fairer than hunting with a rifle, because you have to get so close to the animal to get a good shot. Thirty, thirty-five yards, max. Close enough to hear water sloshing in the elk's belly, from where he's just taken a drink from the creek. Close enough to hear the intakes of breath. Close enough to be fair. But Suzie doesn't see it that way. She'll serve Randy his drinks and chat with him, be polite, but her face is blank, her smiles stiff.

Last summer Randy tried to gain Suzie's favor by building her things. Davey, the bartender — the man she was with at the time — didn't really mind. It wasn't as if there were any threat of Randy stealing her away, and besides, Davey liked the objects Randy built her. And, too, it might have added a small bit of white heat to Davey and Suzie's relationship, though I cannot say for sure.

Randy made her a porch swing out of bright larch wood and stained it with tung oil. He gave it to her at the saloon one night after spending a week on it, sanding it and getting it just right. We gathered around and admired it, running our hands over its smoothness. Suzie smiled a little — a polite smile, which was, in a way, worse than if she had looked angry — and said nothing, not even thank you, and she and Davey took it home in the back of Davey's truck.

Randy built her other things too, small things she could fit on her dresser — a little mahogany box for her earrings, of which she had several pairs — and a walking stick with a deer antler for the grip. She said she didn't want the walking stick but would take the earring box.

That summer I lay awake in my cabin some nights and thought about how Suzie was with Davey. I felt vaguely sorry for him, because I knew she would leave him too. I'd lie on my side and look out my bedroom window at the northern lights flashing. The river runs by my cabin, and the strange flashing reflected on the river in a way that made it seem that the light was coming from beneath the water as well. On nights like those, it felt as if my heart would never heal — in fact, I was certain of it. By then I didn't love Suzie anymore — at least I didn't think I did — but I wanted to love someone, and to be loved.

In the evenings, back when we'd been together, Suzie and I would sit on the porch after she got in from work. There was still plenty of daylight left, and we'd watch large

herds of deer, their antlers still covered with summer velvet, wade into the cool shadows of the river to bathe, like
ladies. They made delicate splashing sounds as they stepped
into the current. Water fell from their muzzles when they
lifted their heads from drinking. As the sun moved lower,
their bodies grew increasingly indistinct, blurring into shadows. Later, Suzie and I would wrap a single blanket about
ourselves and nap. When we opened our eyes we would
watch for falling stars, and wait until we saw one go
ripping across the sky, shot through all the other stars, and
once in a great while they came close enough for us to
hear the crackle and hiss as they burned.

In early July, Randy, whose house sits in a field up at the
base of the mountains, began practicing with his bow.
Standing in the field at various marked distances — ten,
twenty, thirty, forty yards — he shot arrow after arrow at
a bull's-eye stapled to bales of hay. It was unusual to drive
past then and not see him outside, shirtless, perspiring, his
cheeks flushed. He lived by himself and there was probably
nothing else to do. The bowhunting season began in late
August, months before the shooting season.

It made Suzie furious to see Randy practicing with his
bow and arrows. She circulated a petition in the valley to
ban bowhunting.

But we would have been out doing the same thing if
we'd had the skill, hunting giant elk with bows for the thrill
of it, luring them with calls and rattles to our hiding places
in the dark woods. Provoked, the bulls would rush in,
heads down, their great antlers ripping through the under-
brush and knocking against the overhanging limbs of trees.

If only we could have brought them in close. But we
weren't good enough to do that.

We couldn't sign the petition. Not even Davey could sign it.

"But it's wrong," Suzie said.

"It's personal choice," Davey said. "If you use the meat, and apologize to the spirit right before you do it and right after, if you give thanks, it's all right. It's a man's choice, honey."

If there was one thing Suzie hated, it was that man-woman stuff. She shut her eyes and held them shut as if she were trying to be in some other place. "He's trying to prove something," she said.

"Randy's just doing something he cares about, dear," Davey said.

"He's trying to prove his manhood — to me, to all of us," she said. "He's dangerous."

"No," Davey said, "that's not right. He likes it and hates it both. It fascinates him, is all."

"It's sick," she said. "He's dangerous."

I could see that Suzie and Davey did not have that long to go.

I went bowhunting with Randy once to see how it was done. I saw him shoot an elk, saw the arrow go in behind the bull's shoulder, where the heart and lungs are hidden. I saw, too, the way the bull looked around in wild-eyed surprise before galloping off through the timber, seemingly uninjured, running hard. We listened for a long time to the *clack-clack* of the aluminum arrow banging against the trees.

Randy was wearing camouflage fatigues. He'd painted his face in stripes, like a tiger's. "Now we sit and wait." Randy seemed confident, not shaky at all, though I was. It had looked like a record bull, massive and dark. I had smelled anger from the bull — fury — when the arrow first snapped into his ribs, and when he lunged away. I didn't believe we'd ever see him again.

After two hours we got up and began to follow the blood trail. There wasn't much of it at first, just a drop or two in the dry leaves, already turning brown and cracking, drops that I would never have seen had Randy not pointed them out. A quarter of a mile down the hill we began to see more of it, a widening stream of blood, until it seemed that surely all of the bull's blood had drained out. We passed two places where the bull had lain down beneath a tree to die, but had then gotten up and moved on. We found him by the creek a half mile away, down in the shadows, his huge antlers rising into a patch of sun and gleaming. The arrow did not seem large enough to have killed him. The creek made a gentle trickling sound.

We sat down beside the elk and admired him, studied him. Randy, who because of the scent did not smoke during the hunting season, at least not until he had his elk, pulled out a pack of cigarettes, shook one out, and lit it.

"I'm not sure why I do it," he said, reading my mind. "I feel kind of bad about it each time I see one like this, but I keep doing it." He shrugged. I listened to the sound of the creek. "I know it's cruel, but I can't help it, I have to do it," he said.

"What do you think it feels like?" Suzie once asked me at the saloon. "What do you think it feels like to run with an arrow in your heart, knowing you're going to die for it?" She was red-faced, self-righteous.

I told her I didn't know, it was just the way it was. I paid for my drink and left.

Late in July, Suzie left Davey, as I'd predicted. It was an amicable separation, and we all had a party down at the saloon to celebrate. An entire deer was roasted in honor of the occasion: Bud Jennings had hit the deer with his truck the night before, coming back from town with supplies. We sat outside in the early evening and ate the steaming meat off paper plates with barbecue sauce and crisp apples from Idaho. The river that dovetailed with the road glinted in the fading light. This was back when Old Terjaney was still alive, and he played his accordion, a sad, sweet sound. We drank beer and told stories.

All this time I'd been uncertain about whether it was right or wrong to hunt if you used the meat and said those prayers. And I'm still not entirely convinced one way or the other. But I have a better picture of what it's like to be the elk or deer. And I understand Suzie a little better too. She was frightened. Fright — sometimes plain fright, even more than terror — is every bit as bad as pain, and maybe worse.

Suzie went home with me that night, after the party. She had made her rounds of the men in the valley, and now she was choosing to come back to me.

"I've got to go somewhere," she said. "I hate being alone. I can't stand to be alone." She slipped her hand into mine as we walked home. Randy and Davey were still sitting at the picnic table, eating slices of venison. The sun hadn't quite set. Ducks flew down the river.

"I guess that's as close to 'I love you' as I'll get," I said.

"I'm serious," she said, twisting my hand. "You don't understand. It's *horrible*. I can't *stand* it. It's not like other people's loneliness. It's worse."

"Why?" I asked.

"No reason," Suzie said. "I'm just scared, jumpy. I can't help it."

"It's okay," I said.

We walked down the road like that, holding hands in the dusk. It was about three miles down the gravel road to my cabin. Suzie knew the way. We heard owls as we walked along the river, and saw lots of deer. Once, I thought I heard some wild sound and turned to look back, but I saw nothing, saw no one.

If Randy can have such white-hot passion for bowhunting, he surely can have just as much heat in his hate. It spooks me the way he doesn't bring Suzie presents anymore in the old hopeful way. The flat looks he gives me could mean anything: they unnerve me.

Sometimes I'm afraid to go in the woods, but I go anyway. I hunt, working along a ridge, moving in and out of the shadows between the forest and the meadow, walking this line at dusk and thinking about Suzie, sometimes, instead of hunting. I'll think about how comfortable I am with her — how gratified — if not actually in love. I move slowly through the woods, trying to be as quiet as I can. There are times now when I feel someone or something is just behind me, following at a distance, and I'll turn around, frightened and angry both, and I won't see anything.

The day before Halloween it began to snow, and it didn't stop for our party the following night. The roof over the saloon groaned under its heavy load. But we all managed to get together for the dance anyway, swirling around the room, pausing to drink or arm wrestle, the antlers tied securely on our heads. We pretended to be deer or elk, as we always do on Halloween, and pawed at the wideboard floor with our boots. Davey and Suzie waltzed in widening circles; she seemed so light and free that I couldn't help but grin. Randy sat drinking beer off in a corner. At one point he smiled. It was a polite smile.

The beer ran out at three in the morning, and we started to gather our things. Those of us who had skied down to the saloon tried to find someone to tow us home. Because Randy and I lived up the same road, Davey offered us both a ride, and Suzie took hold of the tow rope with us.

Davey drove slowly through the storm. The snowflakes were as large as goose feathers. We kept our eyes on the brake lights in front of us, with the snow spiraling into our faces, and concentrated on gripping the rope.

Suzie had had a lot to drink, we all had, and she held the rope with both hands, her deer antlers slightly askew. She began asking Randy about his hunting — not razzing him, as I thought she would, but simply questioning him — things she'd been wondering for a long time, I supposed, but had been too angry to ask.

"What's it like?" Suzie kept wanting to know. "I mean, what's it *really* like?"

We were sliding through the night, holding on to the rope. The snow struck our faces, caking our eyebrows, and it was so cold that it was hard to speak without shivering. "You're too cold-blooded for me," she said when Randy wouldn't answer. "You scare me, mister."

Randy stared straight ahead, his face hard and flat and blank.

"Suzie, honey," I started to say. I had no idea what I was going to say after that — something to defend Randy, perhaps — but I stopped, because Randy turned and looked at me, for just a second, with a fury I could feel as well as see, even in my drunkenness. But then the mask, the polite mask, came back down over him, and we continued down the road in silence, the antlers on our heads bobbing and weaving, a fine target for anyone who might not have understood that we weren't wild animals.

Wejumpka

WHEN WEJUMPKA was eleven, Vern, his father, made me Wejumpka's godfather. Vern's health had been going downhill fast all that fall, although he was only in his fifties.

Vern and I had been playing that ridiculous game of liar's poker, and our final bet obliged me to be Wejumpka's godfather if I called and lost, but if Vern was bluffing, he had to marry the woman he was then seeing, if she would have him. She was barely out of her teens, plump, with a pear-shaped face and orange hair. She had an easy laugh and had had two children already. I agreed to the bet with the hope that she could do something with him — straighten him up, as I knew women could sometimes do.

The woman's father, who was a few years older than me, had attended my high school. We'd played football together; he was a wiry little halfback who'd gotten more wiry since, working on cars in his garage out near the Pearl River. His name was Zachary, and he collected insurance money each spring when the rains brought the river into his garage and on up into his house. Generally, he didn't even move out when it flooded. He'd wade around, doing his chores, making sure all the circuit breakers were off, and wait for the water to go down. When he'd collected the insurance money, he would bury it in secret places.

The rains began in March, and he would sit on the roof of his garage and listen to the weather station, praying for more rain; each foot of water in his shop was worth about ten thousand dollars. It's cruel, but I don't know what his daughter's name was. Worse, I don't think Vern did either. We called her Zachary's girl.

The final bet between Vern and me was made only around midnight, but we'd started drinking at four in the afternoon. It was a serious hand of poker. Vern wasn't bluffing, it turned out, and as I suspected later, he wasn't even drunk. It was a setup. It was as if I'd killed him, like in one of those hunting accidents where a best buddy trips and hits the trigger, shooting his partner. By losing the bet, by assuming responsibility for Wejumpka, I'd given Vern the last go-ahead he needed to let go of everything, let his downward spiral have its way with him.

After that hand, Vern wasn't the same anymore. I played cards with him out of a sense of duty, and that was probably why I got so drunk — I don't handle duty very well. Later that year, when Vern received the old stop-drinking-or-you'll-only-have-one-year-to-live speech from his doctor, and yet didn't stop, not much anyway, I felt naive and stupid for having called an older man's bluff.

About the boy I have won, Wejumpka. When he was eight, he went with his scout troop on a father-and-son camp-out. They roasted marshmallows and sang songs around the campfire. The moon was high and silvery over the lake. Bats chittered and swooped above the water. There was the cool, sweet trill of a screech owl coming from the woods along the shore. Solemnly, the boys gave each other Indian names, written on slips of paper and drawn from a wooden box.

Every other scout soon forgot his name, or was less than flattered by it and threw the paper in the fire. But Wejumpka remembered his; he embraced it. Formerly his name was Montrose. Another thing about the boy I have inherited: he is a hugger, and he's wild about puppies, cats, parrots, guinea pigs — he loves all animals, and other children too, even the mean ones who pushed him down and ran away when he tried to embrace them.

He's always been that way, always holding on. Perhaps when he was in his mother's womb he could feel, as if with some prenatal sonar, the dark shape of his future, of the divorce looming between Vern and his wife, Ann. Perhaps Vern said unkind words to his wife while Wejumpka was forming in her womb. It's also possible that Ann could see into the future, could feel the absence of a thing. Perhaps she held Vern more tightly than ever then, being wise and clairvoyant and scared in her pregnancy, and this affected the unborn child, made him hold on in the same way.

When Wejumpka was six, the year before the divorce, he dressed up as Porky Pig for Halloween. The other children were devils or witches or Green Berets with rubber knives clenched between their teeth, but Wejumpka was Porky Pig, and he went from house to house hugging people when they answered the door. He never asked for candy, not quite understanding that part of it, but instead ran into these strangers' living rooms and latched on to their legs, giving them a tight thigh hug. Vern and Ann were having one of their dinner parties in which they would end up insulting each other in front of the guests, and it was my job to take Wejumpka around to all the nearby houses and bring him safely back.

Vern and Ann had not started their fight, though, by the time we returned. It's possible that they were still a little in love, or thought they were; when they answered the door and saw their own little Porky Pig standing in front of them, they looked at each other and smiled. They had been drinking.

"Trick or treat!" Wejumpka shouted through his plastic mask, hopping up and down. I had tried to explain to him how it worked, that sometimes it was best not to hug. He was overjoyed, after the running chaos of the night, all the hurried darkness, at seeing his mother and father standing in the doorway with the bright lights of the party behind them, all the safe noise.

"Trick or treat!" he shouted again, jumping up and down once more.

Ann frowned and took a step back. "Why, you're *scary*," she said, and Wejumpka stopped hopping and looked at me.

"Whose little boy are *you?*" Vern asked, bending down and peering into the mask. "I don't believe I *know* any little Porky Pig boys," he said, shaking his head. And they closed the door.

"It's me!" Wejumpka screamed, struggling to get out of the hot costume and mask. "It's me! It's your little boy!"

His parents swung the door open, and this time the guests were gathered around it, laughing as Wejumpka flew into Ann's arms, crying. She patted him on the back and made soft comforting sounds.

After the divorce, Vern's sports car stopped running, and he never fixed it; it sat in the small woodshed-garage behind his apartment. Mice built their nests in and around the engine; they nibbled the insulation off the electrical wiring. Birds nested in the rafters of the shed, and the car was soon dappled with what seemed to be their hearty enthusiasm. When Vern does go to work now, he walks. Or Zachary's girl picks him up. Vern slumps in the seat beside her and drinks rum from the bottle, still wearing the suit he wore the day before, and the day before that.

Sometimes I drive Vern over to Zachary's. We'll play cards at a little linoleum-covered table that rocks whenever we lean our elbows on it. Zachary's girl and Vern drink from the same bottle, but Zachary and I drink from jelly glasses.

"Lot of bad shit goin' around," Zachary will say, shaking his head, studying his hand as if it's the first game of cards he has ever played in his life. Zach's girl and Vern giggle, look at each other's hands. At just such an opportunity, Zachary and I start talking about football, talking as if we'll run ourselves into another winning streak, talking and drinking rum with hope and idiocy. Zach's girl and Vern slide to the floor, in a way becoming rum themselves. They land in a tangled heap.

The room grows quiet. By this time the moon might be up and full. Zachary sighs and turns to the window, thinking perhaps about Vern's rotting sports car — Zachary could fix it, maybe, or weld it to the top of a tower he could climb each day after work. He could sit behind the steering wheel and dream that he was a sailor in a crow's nest, peering out at everything, ever mindful of the treasure he had hidden away.

When Wejumpka entered junior high school, he finally stopped hugging people. The authorities made him stop. They told Ann that he couldn't come to school anymore if he continued. He was hugging teachers, students, the custodians, the principal. He was considered a disciplinary problem.

Vern and I decided to change his name then. He was getting too old for Wejumpka — though it's what Vern and I still call him — but God knows, Montrose was nothing to fall back on. In order to settle the matter, we told him to pick a name from a wooden bowl that had a lot of slips of paper in it, and "Vern Jr." was the one he pulled out. Fate.

He is luckier than Zach's girl, though, and luckier even than Zach, who has some malaise in his blood, some unknown chemical that makes him have to lie down and rest every time the wind changes direction.

For his twelfth birthday, I rented a boat and two pairs of water skis and took Wejumpka and his cousin Austin, who was sixteen, out to a nearby lake, just the two boys and I. Neither of the boys had skied before, and for a long time we let our boat idle, feeling the warmth of the sun. Zachary had finally towed Vern's sports car away and had indeed welded it to the top of a tower on the far side of the lake. We were some distance from shore, so we had to use binoculars to see it.

"I pissed in that car after they got the divorce," Austin bragged, proud and tough. He was wearing a gold earring and a dirty blue-jean jacket, even though it was close to one hundred degrees. His body gave off a fetid odor, like a boys' restroom at school, and I wanted him to ski first so that he would get cleaned off.

"I pooted in it," Wejumpka admitted in a small voice. The two cousins looked at each other and then broke up laughing. I laughed too, at the coincidence of this. It came to me then how good it would feel to turn on the engine and go. The water was deep, and I could see a long way down into it, or so it seemed. I noticed fish shilly-shallying and the square marks on a turtle's back.

Wejumpka, in an odd gesture of bravery, asked to ski first. It's possible that he wanted to show off for Austin, or perhaps he thought his father had not yet abandoned him, that he was being given one last chance. Perhaps even as he climbed down into the water, buoyed by his life vest, and slipped his feet into the oversized skis — even then, in his staggered, hugging-poet's imagination, his father was climbing up into the car with Zachary, watching him through binoculars, giving him a final chance, maybe even elbowing Zachary and pointing to his son, saying, "That's my little boy. That's my Wejumpka."

I started the motor and tipped our craft from side to side, getting the tow rope lined up, making sure Wejumpka had his skis on properly. We moved forward slowly, but he lost his balance and went down. He was stout, though, and he bobbed right up, with a surprised look on his face. We tried again.

When I glanced over my shoulder, I saw that he wasn't watching the boat but the shore, squinting as if he were waiting for something to appear.

"He says he wants to go faster!" Austin shouted, amazed at the spectacle of his cousin. "He's pointing his thumb up. He wants to go faster!"

I turned again and saw that it was so. Wejumpka was leaning back like a pro. Already he was relaxed, with a cocky but determined look on his face, and he was jabbing his thumb at the sky as if trying to poke out the bottom of something, skiing with just one hand.

I eased the throttle in. The boat surged forward like a lion, but Wejumpka would not allow himself to be easily left behind. I saw that he was a little pale — the throttle was now all the way in. He was crouching into the wake, no longer showboating, trying only to hang on. We neared a wall of blue trees, and almost without realizing it, I noticed that we could see without effort the car on top of its tower, looking like the most natural thing in the world.

We skimmed the chop of summer-wind waves. The breeze was blowing my hair, and the sun was beginning to burn my cheeks and shoulders. When I looked back, Wejumpka was gone. Austin was staring open-mouthed at the water behind us.

Then we saw that he was still holding on to the end of the rope, though the skis were knocked off by his fall; he was bright as a fishing lure. Occasionally he raised his head above the rooster tail of water, his mouth a tiny, frightened *O*, gulping for air. The force of the water must have been tremendous.

"Let go!" I shouted, easing back on the throttle. "Wejumpka, *let go!*" I could feel the strain on the boat.

But he couldn't hear me underwater. I had to shut the engine off and coast to a stop before he understood that the ride was over.

The Legend
of Pig-Eye

WE USED TO GO to bars, the really seedy ones, to find our fights. It excited Don. He loved going into the dark old dives, ducking under the doorway and following me in, me with my robe on, my boxing gloves tied around my neck, and all the customers in the bar turning on their stools, as if someday someone special might be walking in, someone who could even help them out. But Don and I were not there to help them out.

Don had always trained his fighters this way: in dimly lit bars, with a hostile hometown crowd. We would get in his old red truck on Friday afternoons — Don and Betty, his wife, and Jason, their teenage son, and my two hounds, Homer and Ann — and head for the coast — Biloxi, Ocean Springs, Pascagoula — or the woods, to the Wagon Wheel in Utica. If enough time had passed for the men to have forgotten the speed of the punches, the force and snap of them, we'd go into Jackson, to the rotting, sawdust-floor bars like the Body Shop or the Tall Low Man. That was where the most money could be made, and it was sometimes where the best fighters could be found.

Jason waited in the truck with the dogs. Occasionally Betty would wait with him, with the windows rolled down so they could tell how the fight was going. But there were times when she went with us into the bar, because that raised the stakes: a woman, who was there only for the fight. We'd make anywhere from five hundred to a thousand dollars a night.

"Mack'll fight anybody, of any size or any age, man or woman," Don would say, standing behind the bar with his notepad, taking bets, though of course I never fought a woman. The people in the bar would pick their best fighter, and then watch that fighter, or Betty, or Don. Strangely, they never paid much attention to me. Don kept a set of gloves looped around his neck as he collected the bets. I would look around, wish for better lighting, and then I'd take my robe off. I'd have my gold trunks on underneath. A few customers, drunk or sober, would begin to realize that they had done the wrong thing. But by that time things were in motion, the bets had already been made, and there was nothing to do but play it out.

Don said that when I had won a hundred bar fights I could go to New York. He knew a promoter there to whom he sent his best fighters. Don, who was forty-four, trained only one fighter at a time. He himself hadn't boxed in twenty years. Betty had made him promise, swear on all sorts of things, to stop once they got married. He had been very good, but he'd started seeing double after one fight, a fight he'd won but had been knocked down in three times, and he still saw double, twenty years later, whenever he got tired.

We'd leave the bar with the money tucked into a cigar

box. In the summer there might be fog or a light mist falling, and Don would hold my robe over Betty's head to keep her dry as we hurried away. We used the old beat-up truck so that when the drunks, angry that their fighter had lost, came out to the parking lot, throwing bottles and rocks at us as we drove away, it wouldn't matter too much if they hit it.

Whenever we talked about the fights, after they were over, Don always used words like "us," "we," and "ours." My parents thought fighting was the worst thing a person could do, and so I liked the way Don said "we": it sounded as though I wasn't misbehaving all by myself.

"How'd it go?" Jason would ask.

"We smoked 'em," Don would say. "We had a straight counterpuncher, a good man, but we kept our gloves up, worked on his body, and then got him with an overhand right. He didn't know what hit him. When he came to, he wanted to check our gloves to see if we had put *lead* in them."

Jason would squeal, smack his forehead, and wish that he'd been old enough to see the bout for himself.

We'd put the dogs, black-and-tan pups, in the back of the truck. The faithful Homer, frantic at having been separated from me, usually scrambled around, howling and pawing; but fat Ann curled up on a burlap sack and quickly fell asleep. We'd go out for pizza then, or to a drive-through hamburger place, and we'd talk about the fight as we waited for our order. We counted the money to make sure it was all there, though if it wasn't, we sure weren't going back after it.

We could tell just by looking at the outside what a place was going to be like, if it was the kind of place where we would have to leave Betty in the truck with Jason, sometimes with the engine running, and where we didn't know for sure if we would win or lose.

We looked for the backwoods night spots, more gathering places than bars, which were frequented by huge, angry men — men who either worked hard for a living and hated their jobs or did not work and hated that too, or who hated everything, usually beginning with some small incident a long time ago. These were the kinds of men we wanted to find, because they presented as much of a challenge as did any pro fighter.

Some nights we didn't find the right kind of bar until almost midnight, and during the lull Betty would fall asleep with her head in Don's lap and Jason would drive so I could rest; the dogs curled up on the floorboard. Finally, though, there would be the glow of lights in the fog, the crunch of a crushed-shell parking lot beneath our tires, and the cinder-block tavern, sometimes near the Alabama state line and set back in the woods, with loud music coming through the doors, seeping through the roof and into the night; between songs we could hear the clack of pool balls. When we went in the front door, the noise would come upon us like a wild dog. It was a furious caged sound, and we'd feel a little fear in our hearts. Hostility, the smell of beer, and anger would swallow us up. It would be just perfect.

"We'll be out in a while," Don would tell Jason. "Pistol's in the glove box. Leave the engine running. Watch after your mother."

We kept a tag hanging from the truck's rear-view mirror that told us how many fights in a row we had won, what the magic number was, and after each fight it was Jason's job to take down the old tag and put a new one up.

Eighty-six. Eighty-seven. Eighty-eight.

• • •

Driving home, back to Don's little farm in the woods, Jason would turn the radio on and steer the truck with one hand, keeping the other arm on the seat beside him, like a farmer driving into town on a Saturday. He was a good driver. We kept rocking chairs in the back of the truck for the long drives, and sometimes after a fight Don and I would lean back in them and look up at the stars and the tops of big trees that formed tunnels over the lonely back roads. We'd whistle down the road as Jason drove hard, with the windows down and his mother asleep in the front.

When a road dipped down into a creek bottom, the fog made it hard to see beyond the short beam of our headlights, as if we were underwater. The air was warm and sticky. Here Jason slowed slightly, but soon we'd be going fast again, driving sixty, seventy miles an hour into the hills, where the air was clean and cool, and the stars visible once more.

I wondered what it would be like to drive my father and mother around like that, to be able to do something for them, something right. My parents lived in Chickasay, Oklahoma, and raised cattle and owned a store. I was twenty years old.

I wanted to win the one hundred fights and go to New York and turn pro and send my parents money. Don got to keep all of the bar-fight money, and he was going to get to keep a quarter of the New York money, if there ever was any. I wanted to buy my parents a new house or some more cattle or something, the way I read other athletes did once they made it big. My childhood had been wonderful; already I was beginning to miss it, and I wanted to give them something in return.

When I took the robe off and moved in on the bar fighter, there was Don and Betty and Jason to think of too. They were just making expenses, nothing more. I could not bear to think of letting them down. I did not know what my parents wanted from me, but I did know what Don and Jason and Betty wanted, so that made it easier, and after a while it became easier to pretend that it was all the same, that everyone wanted the same thing, and all I had to do was go out there and fight.

Don had been a chemist once for the coroner's lab in Jackson. He knew about chemicals, drugs. He knew how to dope my blood, days before a fight, so I would feel clean and strong, a new man. He knew how to give me smelling salts, sniffs of ammonia vials broken under my nose when I was fighting sloppily, sniffs that made my eyes water and my nose and lungs burn, but it focused me. And even in training, Don would sometimes feint and spar without gloves and catch me off guard, going one way when I should have been going the other. He would slip in and clasp a chloroform handkerchief over my face. I'd see a mixed field of black and sparkling, night-rushing stars, and then I'd be down, collapsed in the pine needles by the lake where we did our sparring. I'd feel a delicious sense of rest, lying there, and I'd want to stay down forever, but I'd hear Don shouting, ". . . Three! Four! Five!" and I'd have to roll over, get my feet beneath me, and rise, stagger-kneed, the lake a hard glimmer of heat all around me. Don would be dancing around me like a demon, moving in and slapping me with that tremendous reach of his and then dancing back. I had to get my gloves up and stay up, had to follow the blur of him with that backdrop of deep woods and lake, with everything looking new and different suddenly, making no sense; and that, Don said, was what it was like to get knocked out. He wanted me to practice it occasionally, so that I would know what to do when it finally happened, in New York, or Philadelphia, or even in a bar.

My body hair was shaved before each fight. I'd sit on a chair by the lake in my shorts while the three of them, with razors and buckets of soapy water, shaved my legs, back, chest, and arms so the blows would slide away from me rather than cut in, and so I would move faster, or at least *feel* faster — that new feeling, the feeling of being some-one else, newer, younger, and with a fresher start.

When they had me all shaved, I would walk out on the dock and dive into the lake, plunging deep, ripping the water with my new slipperiness. I would swim a few easy strokes out to the middle, where I would tread water, feeling how unbelievably smooth I was, how free and unattached, and then I would swim back in. Some days, walking with Don and Betty and Jason back to the house, my hair slicked back and dripping, with the woods smelling good in the summer and the pine needles dry and warm beneath my bare feet — some days, then, with the lake behind me, and feeling changed, I could almost tell what it was that everyone wanted, which was nothing, and I was very happy.

After our bar fights, we'd get home around two or three in the morning. I'd nap on the way, in the rocker in the back of the truck, rocking slightly, pleasantly, whenever we hit a bump. Between bumps I would half dream, with my robe wrapped tightly around me and the wind whipping my hair, relaxed dreams, cleansed dreams — but whenever I woke up and looked at Don, he would be awake.

He'd be looking back at where we'd come from, the stars spread out behind us, the trees sliding behind our taillights, filling in behind us as if sealing off the road. Jason would be driving like a bat out of hell, with the windows down, and coffee cups and gum wrappers swirled around in the truck's cab.

Sometimes Don turned his chair around to face the cab, looking in over Jason's shoulder and watching him drive, watching his wife sleep. Don had been a good boxer but the headaches and double vision had gotten too bad. I wondered what it would take for me to stop. I could not imagine anything would. It was the only thing I could do well.

On the long, narrow gravel road leading to Don's farm, with the smell of honeysuckle and the calls of chuck-will's-widows, Jason slowed the truck and drove carefully, respecting the value of home and the sanctity of the place. At the crest of the hill he turned the engine and lights off and coasted the rest of the way to the house, pumping the squeaky brakes, and in silence we'd glide down the hill. From here the dogs could smell the lake. They scrambled to their feet, leapt over the sides of the truck, and raced to the water to inspect it and hunt for frogs.

I slept in a little bunkhouse by the lake, a guest cottage they had built for their boxers. Their own place was up on the hill. It had a picnic table out front and a garage — it was a regular-looking house, a cabin. But I liked my cottage. I didn't even have a phone. I had stopped telling my parents about the fights. There was not much else to tell them about other than the fights, but I tried to think of things that might interest them. Nights, after Don and Betty and Jason had gone to bed, I liked to swim to the middle of the lake, and with the moon burning bright above me, almost like a sun, I'd float on my back and fill my lungs with air. I'd float there for a long time.

The dogs swam around me, loyal and panting, paddling in frantic but determined circles, sneezing water. I could feel the changing currents beneath and around me as the dogs stirred the water, and could see the wakes they made, glistening beneath the moon — oily-black and mint-white swirls. I loved the way they stayed with me, not knowing how to float and instead always paddling. I felt like I was their father or mother. I felt strangely like an old man, but with a young man's health.

I'd float like that until I felt ready again, until I felt as if I'd never won a fight in my life — in fact, as if I'd never even fought one, as if it was all new and I was just starting out and had everything still to prove. I floated there until I believed that that was how it really was.

I was free then, and I would break for shore, swimming again in long, slow strokes. I'd get out and walk through the trees to my cottage with the dogs following, shaking water from their coats and rattling their collars, and I knew the air felt as cool on them as it did on me. We couldn't see the stars, down in the trees like that, and it felt very safe.

I'd walk through the woods, born again in my love for a thing, the hard passion of it, and I'd snap on my yellow porch light as I went into the cottage. The light seemed to pull in every moth in the county. Homer and Ann would stand on their hind legs and dance, snapping at the moths. Down at the lake the bullfrogs drummed all night, and from the woods came the sound of crickets and ka-tydids. The noise was like that at a baseball game on a hot day, always some insistent noises above others, rising and falling. I could hear the dogs crunching June bugs as they caught them.

Right before daylight Betty would ring a bell to wake me for breakfast. Don and I ate at the picnic table, a light breakfast, because we were about to run, me on foot and Don on horseback.

"You'll miss me when you get up to New York," he said. "They'll lock you in a gym and work on your technique. You'll never see the light of day. But you'll have to do it."

I did not want to leave Betty and Jason, did not even want to leave Don, despite the tough training sessions. It would be fun to fight in a real ring, with paying spectators, a canvas mat, a referee, and ropes, safety ropes to hold you in. I would not mind leaving the bar fights behind at all, but I could not tell Don about my fears. I was half horrified that a hundred wins in Mississippi would mean nothing, and that I would be unable to win even one fight in New York.

Don said I was "a fighter, not a boxer."

He'd had other fighters who had gone on to New York, who had done well, who had won many fights. One of them, his best before me, Pig-Eye Reeves, had been ranked as high as fifth as a WBA heavyweight. Pig-Eye was a legend, and everywhere in Mississippi tales were told about him. Don knew all of them.

Pig-Eye had swum in the lake I swam in, ate at the same picnic table, lived in my cottage. Pig-Eye had run the trails I ran daily, the ones Don chased me down, riding his big black stallion, Killer, and cracking his bullwhip.

That was how we trained. After breakfast Don headed for the barn to saddle Killer, and I whistled the dogs up and started down toward the lake. The sun would be coming up on the other side of the woods, burning steam and mist off the lake, and the air slowly got clearer. I could pick out individual trees through the mist on the far side. I'd be walking, feeling good and healthy, at least briefly, as if I would never let anyone down. Then I would hear the horse running down the hill through the trees, coming after me, snorting, and I'd hear his hooves and the saddle creaking, with Don riding silently, posting. When he spotted me, he'd crack the whip once — that short mean *pop!* — and I would have to run.

Don made me wear leg weights and wrist weights. The dogs, running beside me, thought it was a game. It was not. For punishment, when I didn't run fast enough and Killer got too close to me, Don caught my shoulder with the tip of the whip. It cut a small strip into my sweaty back, which I could feel in the form of heat. I knew this meant nothing, because he was only doing it to protect me, to make me run faster, to keep me from being trampled by the horse.

Don wore spurs, big Mexican rowels he'd bought in an antiques store, and he rode Killer hard. I left the trail sometimes, jumping over logs and dodging around trees and reversing my direction, but still Killer stayed with me, leaping the same logs, galloping through the same brush, though I was better at turning corners and could stay ahead of him that way.

This would go on for an hour or so, until the sun was over the trees and the sky bright and warm. When Don figured the horse was getting too tired, too bloody from the spurs, he would shout "Swim!" and that meant it was over, and I could go into the lake.

"The Lake of Peace!" Don roared, snapping the whip and spurring Killer, and the dogs and I splashed out into the shallows. I ran awkwardly, high-stepping the way you do going into the waves at the beach. I leaned forward and dropped into the warm water, felt the weeds brushing my knees. Killer was right behind us, still coming, but we would be swimming hard, the dogs whining and rolling their eyes back like Chinese dragons, paddling furiously, trying to see behind them. By now Killer was swimming too, blowing hard through his nostrils and grunting, much too close to us, trying to swim right over the top of us, but the dogs stayed with me, as if they thought they could protect me, and with the leg weights trying to weigh me down and pull me under, I'd near the deepest part of the lake, where the water turned cold.

I swam to the dark cold center, and that was where the horse, frightened, slowed down, panicking at the water's coldness and swimming in circles rather than pushing on. The chase was forgotten then, but the dogs and I kept swimming, with the other side of the lake drawing closer at last, and Jason and Betty standing on the shore, jumping and cheering. The water began to get shallow again, and I came crawling out of the lake. Betty handed me a towel, Jason dried off the dogs, and then we walked up the hill to the cabin for lunch, which was spread out on a checkered tablecloth and waiting for me as if there had never been any doubt that I would make it.

Don would still be laboring in the water, shouting and cursing at the horse now, cracking the whip and giving him muted, underwater jabs with his spurs, trying to rein Killer out of the angry, confused circles he was still swimming, until finally, with his last breath, Killer recognized that the far shore was as good as the near one, and they'd make it in, struggling, twenty or thirty minutes behind the dogs and me.

Killer would lie on his side, gasping, coughing up weeds, ribs rising and falling, and Don would come up the hill to join us for lunch: fried chicken, cream gravy, hot biscuits with honey, string beans from the garden, great wet chunks of watermelon, and a pitcher of iced tea for each of us. We ate shirtless, barefoot, and threw the rinds to the dogs, who wrestled and fought over them like wolves.

At straight-up noon, the sun would press down through the trees, glinting off the Lake of Peace. We'd change into bathing suits, all of us, and inflate air mattresses and carry them down to the lake. We'd wade in up to our chests and float in the sun, our arms trailing loosely in the water. We'd nap as if stunned after the heavy meal, while the dogs whined and paced the shore, afraid we might not come back.

Killer, still lying on the shore, would stare glassy-eyed at nothing, ribs still heaving. He would stay like that until mid-afternoon, when he would finally roll over and get to his feet, and then he would trot up the hill as if nothing had happened.

We drifted all over the lake in our half stupor, our sated summer-day sleep. My parents wanted me to come home and take over the hardware store. But there was nothing in the world that could make me stop fighting. I wished that there was, because I liked the store, but that was simply how it was. I felt that if I could not fight, I might stop breathing, or I might go down: I imagined that it was like drowning, like floating in the lake, and then exhaling all my air, and sinking, and never being heard from again. I could not see myself ever giving up fighting, and I wondered how Don had done it.

We floated and lazed, dreaming, each of us spinning out in different directions whenever a small breeze blew, eventually drifting farther and farther apart, but on the shore the dogs followed only me, tracking me around the lake, staying with me, whining for me to come back to shore.

On these afternoons, following an especially good run and an exhausting swim, I would be unable to lift my arms. Nothing mattered in those suspended, floating times. This is how I can give up, I'd think. This is how I can never fight again. I can drop out, raise a family, and float in the bright sun all day, on the Lake of Peace. This is how I can do it, I'd think. Perhaps my son could be a boxer.

Fights eighty-nine, ninety, ninety-one: I tore a guy's jaw off in the Body Shop. I felt it give way and then detach, heard the ripping sound as if it came from somewhere else, and it was sickening — we left without any of the betting money, gave it all to his family for the hospital bill, but it certainly did not stop me from fighting, or even from hitting hard. I was very angry about something, but did not know what. I'd sit in the back of the truck on the rides home, and I'd know I wanted something, but did not know what.

Sometimes Don had to lean forward and massage his temples, his head hurt so bad. He ate handfuls of aspirin, ate them like M&M's, chasing them down with beer. I panicked when he did that, and thought he was dying. I wondered if that was where my anger came from, if I fought so wildly and viciously in an attempt, somehow and with no logic, to keep things from changing.

On the nights we didn't have a fight, we would spar a little in the barn. Killer watched us wild-eyed from his stall, waiting to get to me. Don made me throw a bucket of lake water on him each time I went into the barn, to make sure that his hate for me did not wane. Killer screamed whenever I did this, and Jason howled and blew into a noise-maker and banged two garbage can lids together, a deafening sound inside the barn. Killer screamed and reared on his hind legs and tried to break free. After sparring we went into the house, and Betty fixed us supper.

We had grilled corn from Betty's garden and a huge porterhouse steak from a steer Don had slaughtered himself, and Lima beans and Irish potatoes, also from the garden. It felt like I was family. We ate at the picnic table as fog moved in from the woods, making the lake steamy. It was as if everyone could see what I was thinking then; my thoughts were bare and exposed, but it didn't matter, because Don and Betty and Jason cared for me, and also because I was not going to fail.

After dinner we watched old fight films. For a screen we used a bedsheet strung between two pine trees. Don set up the projector on the picnic table and used a crooked branch for a pointer. Some of the films were of past champions, but some were old movies of Don fighting. He could make the film go in slow motion, to show the combinations that led to knockdowns, and Betty always got up and left whenever we watched one of the old splintery films of Don's fights. It wasn't any fun for her, even though she knew he was going to win, or was going to get up again after going down.

I had seen all of Don's fights a hundred times and had watched all the films of the greatest fighters a thousand times, it seemed, and I was bored with it. Fighting is not films, it's experience. I knew what to do and when to do it. I'd look past the bedsheet, past the flickering washes of light, while Jason and Don leaned forward, breathless, watching young Don stalk his victim, everything silent except for the clicking of the projector, the crickets, the frogs, and sometimes the owls. In the dark I wondered what New York was going to be like, if it was going to be anything like this.

Some nights, after the movies had ended, we would

talk about Pig-Eye Reeves. It had been several years ago, but even Jason remembered him. We were so familiar with the stories that it seemed to all of us — even to me, who had never met him — that we remembered him clearly.

Pig-Eye knocked out one of the fighters Don had trained, in a bar up in the Delta one night, the Green Frog. That was how Don found Pig-Eye — he had beaten Don's challenger, had just stepped up out of the crowd. Don's fighter, whose name Don always pretended he couldn't remember, threw the first punch, a wicked, winging right, not even bothering to set it up with a jab — Don says he covered his face with his hands and groaned, knowing what was going to happen. Pig-Eye, full of beer, was still able to duck it, evidently, because Don heard nothing but the rip of air and then, a little delayed, the sound of another glove hitting a nose, then a grunt, and the sound of a body falling in the sawdust.

Don and Jason and Betty left the semiconscious fighter there in the Green Frog, with a broken nose and blood all over his chest and trunks. They drove home with no money and Pig-Eye.

They changed the number on the truck mirror from whatever it had been before — forty-five or fifty — back to one. Pig-Eye had won one fight.

"You just left your other fighter sitting there?" I asked the first time I heard the story, though I knew better than to ask now.

Don had seemed confused by the question. "He wasn't my fighter anymore," he said finally.

Sometimes Jason would ask the question for me, so I didn't have to, and I could pretend it didn't matter, as if I weren't even thinking about it.

"Is Mack a better fighter than Pig-Eye?" he'd ask after watching the movies.

Don answered like a trainer every time. He was wonderful, the best. "Mack is better than Pig-Eye ever dreamed of being," he'd say, clapping a big hand on my neck and giving it the death squeeze, his hand the size of a license plate.

"Tell him about the balloon," Jason would cry when Don had reached a fever pitch for Pig-Eye stories.

Don leaned back against a tree and smiled at his son. The lights were off in the house. Betty had gone to bed. Moths fluttered around the porch light, and down below us in the Lake of Peace, bullfrogs drummed. There was no other sound.

"Pig-Eye won his last five fights down here with one hand tied behind his back," Don said, closing his eyes. I wondered if I could do that, wondered if in fact I'd *have* to do that, to ride down the legend of Pig-Eye, and pass over it.

"We sent him up to New York, to a promoter I knew" — Don looked at me quickly — "the same one we'll be sending Mack to if he wins the rest of his fights. This promoter, Big Al Wilson, set him up in a penthouse in Manhattan, had all Pig-Eye's meals catered to him. He had masseurs, everything. He was the *champ.* Everyone was excited about him."

"Tell him about the scars," Jason said. He moved next to his dad, so that his back was against the same tree, and it was as if they were both telling me the story now, though I knew it already, we all knew it.

"Pig-Eye had all these scars from his bar fights," Don said. "He'd been in Vietnam too, and had got wounded there. He flew those crazy hot-air balloons for a hobby, once he started winning some fights and making some money, and he was always having rough landings, always crashing the balloons and getting cut up that way."

"Helium balloons," Jason said.

"It was a very disturbing thing to Pig-Eye's opponents when he first stepped in the ring against them. They'd all heard about him, but he really had to be seen to be believed."

"Like a zipper," Jason said sleepily, but delighted. "He looked like a zipper. I remember."

"Pig-Eye won fourteen fights in New York. He was ranked fifth and was fighting well. I went to a few of his fights, but then he changed."

"He got different," Jason cautioned.

"He stopped calling, stopped writing, and he started getting a little fat, a little slow. No one else could tell it, but I could."

"He needed Dad for a trainer," said Jason. In the distance I heard Killer nicker in his stall.

"He lost," Don said, shaking his head. "He was fighting a nobody, some kid from Japan, and that night he just didn't have it. He got knocked down three times. I saw tapes of it later. He was sitting up like one of those bears in a zoo, still trying to get on his feet for a third time, but he couldn't do it. It was like he didn't know where his legs were, didn't know what his feet were for. He couldn't remember how to do it."

I thought about the ammonia and the chloroform handkerchiefs Don would sometimes place over my face when we were sparring. I wondered if every time he did that to me, he was remembering how Pig-Eye couldn't stand up — how he had forgotten how to get back up. I thought that I surely knew how Pig-Eye had felt.

"The balloon," Jason said. There was a wind in the trees, many nights, and so often those winds reminded me of that strange feeling of being both old and young, someplace in the middle, and for the first time, with no turning back.

"The balloon," Jason said again, punching his father on the shoulder. "This is the best part."

"Pig-Eye was crushed," Don said, sleepy, detached, as if it were no longer Pig-Eye he was talking about. I thought again of how they had walked off and left that other fighter up in the Delta, the nameless one, sitting in the sawdust holding his broken nose. "It was the only time Pig-Eye had ever been knocked out, the only time he'd ever lost, and it devastated him."

"A hundred and fifteen fights," Jason said, "and he'd only lost one."

"But it was my fault," Don said. "It was how I trained him. It was wrong."

"The balloon," Jason said.

"He rented one," Don said, looking up at the stars, speaking to the night. "He went out over the countryside the next day, his face all bandaged up, with a bottle of wine and his girlfriend, and then he took it up as high as it could go, and then he cut the strings to the gondola."

"He was good," Jason said solemnly.

"He was too good," Don said.

All that summer I trained hard for New York. I knew that I would win my hundred fights. I knew that I could win them with one arm tied behind my back, either arm, if Don and Jason wanted that. But I wasn't worried about my one hundred bar fights. I was worried about going up to New York, to a strange place, someplace different. Sometimes I did not want to fight anymore, but I never let anyone see that.

Jason was getting older, filling out, and sometimes Don let him ride Killer. We'd all have breakfast as usual, then Jason would saddle Killer. I'd wake the dogs and we'd start down toward the lake, moving lazily through the trees but knowing that in a minute or two we'd be running.

Don would sit in a chair by the shore and follow us with his binoculars. He had a whistle he'd blow to warn me when I was about to be trampled.

When the dogs and I heard the horse, the hard, fast hooves coming straight down the hill, we'd start to run. It would be almost six o'clock then. The sun would just be coming up, and we'd see things as we raced through the woods: deer slipping back into the trees, cottontails diving into the brush. The dogs would break off and chase all of these things, and sometimes they'd rejoin me later on the other side of the lake with a rabbit hanging from their jaws. They'd fight over it, really wrestling and growling.

All of this would be going past at what seemed like ninety miles an hour: trees, vines, logs; greens, browns, blacks, and blues — flashes of the lake, flashes of sky, flashes of logs on the trail. I knew the course well, knew when to jump, when to dodge. It's said that a healthy man can outrun a horse, over enough distance, but that first mile was the hardest, all that dodging.

Jason shouted, imitating his father, cracking the whip; the sun rose orange over the tops of the trees, the start of another day of perfection. And then the cry, "The Lake of Peace!" And it would be over, and I'd rush out into the shallows, a dog on either side of me, tripping and falling, the lake at my ankles, at my knees, coming up around my waist, and we'd be swimming, with Killer plunging in after us, and Jason still cracking the whip.

· · ·

Actually, there were two stories about Pig-Eye Reeves. I was the only person Don told about the second one. I did not know which one was true.

In the other story, Pig-Eye recovered, survived. Still distraught over losing, he went south, tried to go back to Don, to start all over again. But Don had already taken on another fighter and would not train Pig-Eye anymore.

Don rubs his temples when he tells me this. He is not sure if this is how it went or not.

So Pig-Eye despaired even more and began drinking bottles of wine, sitting out on the dock and drinking them down the way a thirsty man might drink water. He drank far into the night, singing at the top of his lungs. Don and Betty had to put pillows over their heads to get to sleep, after first locking the doors.

Then Don woke up around midnight — he never could sleep through the night — and he heard splashing. He went outside and saw that Pig-Eye had on his wrist and ankle weights and was swimming out to the middle of the lake.

Don said he could see Pig-Eye's wake, could see Pig-Eye at the end of it, stretching it out, splitting the lake in two — and then he disappeared. The lake became smooth again.

Don said that he sleepwalked, and thought perhaps what he'd seen wasn't real. They had the sheriff's department come out and drag the lake, but the body was never found. Perhaps he was still down there, and would be forever.

Sometimes, as Jason and the horse chased me across the lake, I would think about a game I used to play as a child, in the small town in Oklahoma where I grew up.

When I was in the municipal swimming pool, I would hold my breath, pinch my nose, duck under the water, and shove off from the pale blue side of the pool. Like a frog breast-stroking, eyes wide and reddening from the chlorine, I would try to make it all the way to the other side without having to come up for air.

That was the trick, to get all the way to the other side. Halfway across, as the water deepened, there'd be a pounding in the back of my head, and a sinister whine in my ears, my heart and throat clenching.

I thought about that game, as I swam with Jason and Killer close behind me. I seemed to remember my dogs being with me then, swimming in front of me, as if trying to show me the way, half pulling me across. But it was not that way at all, because this was many years before their time. I knew nothing then about dogs, or boxing, or living, or of trying to hold on to a thing you loved, and letting go of other things to do it.

I only understood what it was like to swim through deeper and deeper water, trying as hard as I could to keep from losing my breath, and trying, still, to make it to the deep end.

The Wait

WE DRIVE through the city, through the rain, January: a man I've never met before, Jack, and my best friend, Kirby, still my best friend after twenty years. Jack and Kirby live in the city and are practically best friends themselves now. A dentist and a real estate appraiser. I drove three days and nights to go fishing with them — not just wade fishing, not sissy-pants shore fishing, but in a boat. Jack has a boat with a motor and everything.

I watch Jack as he drives. He looks serious, intent. He's poor, even though he's a dentist, because he's got a wife and three kids, and because this is not a good time in Houston, for dentists or anyone else. Jack's boat is old, and the chances are good that something will go wrong with it today — if we even get out on the water. Kirby is not so poor. Both he and his wife work, and they have only one child, a little girl, who is also named Kirby.

We drive slowly through the thunder and lightning in downtown Houston. Tall buildings leap into the sky all around us with each lightning flash. We pass the building where Kirby works; we pass the building where his wife works. They look like high-rise jails to me, the shutdown of a life. I feel like an outlaw sitting in the back of the jeep, riding with two married men, the fathers, up front. I feel almost as if one of them *is* my father, and the other one of my father's friends. In fact, I am a little older than both of them. It doesn't help that I have never fished from a boat before.

It has been so long since I've been around anyone, man or woman, other than my girlfriend. We've separated. We have done this before, and I think we'll get back together again, because we've been together far too long *not* to get back together.

This time, after Margie left, it was a little different. I felt alone right away, and also, I wanted to do something new, and I did not want to be in that house alone.

It wasn't the usual list of griefs this time — not, Why don't we get married? not, Why are you always traveling so much? not, Who was that woman who called? Those are the little things, the things that can be erased. Or if not erased, at least put aside.

This time, Margie said, she was tired. Just tired. A little frightened, but mostly tired. She went home, back to Virginia.

I did not want to be in that house alone. I just wanted something new. And after a while, that gets hard to find.

We listen to the crackle of the local AM station, the early morning fishing report. The roads are slick, and there are other cars out, so many other cars, but none of them are pulling boats.

Kirby and Jack lean forward and watch the road and sip their coffee. The rain is coming straight down, beating against the windshield. We all try to hear what the Fish Man is saying on his talk show. He is telling us the fishing has been poor to spotty the past few days. Kirby grins, but Jack scowls and says, "Got-damn."

I don't really care one way or the other.

In Texas young men and women are taught to believe the world can be tamed. It's a bull that can be wrestled, and with strength and courage and energy you can lift that bull over your head, spin it around, and throw it to the ground. In certain parts of the world, and even in certain individuals, such a thought would be ludicrous. But in Texas I have seen the myth become truth, lightning strikes, men and women burning across the prairie of their lives, living fast, living strong. I have *seen* it, in my father, my mother, and others, and I feel like an imposter, not having any children to follow after me, even though I am trying to live one of those strong lives myself — fast and free, scorning weakness.

A bolt of lightning smashes down on our left, tingles the hair on our arms. Jack shouts in his fear, and Kirby laughs, leans back in his seat, and rolls the window down a little. A few flicks of rain blow in on his face, and on mine in the back seat. It feels good, and I crack my window a little too.

I mean, Margie takes care of me. Sometimes I get really wild, just run out of the house and up onto the mountain behind our cabin, just *running*. I'll be gone all afternoon, lying up there on some damn rock or something, like a dog in the high mountain sun. When I finally come back, late in the day, she'll be very quiet, and we'll sit together and all will be real calm. What I'm saying is, she takes care of me. And I take care of her, I do. But it's not enough, I think.

Not only are Jack and Kirby best friends now, but their wives are too.

"I dreamed you were in my garbage can last night," Kirby tells Jack matter-of-factly. "I dreamed you were a raccoon, banging around in the garbage, sorting through my trash."

"Uh-huh," says Jack, seemingly amused at the thought of being taken for a raccoon.

There's a metal box in the back of the jeep, a strange-looking box with small holes punched in the sides of it, and I keep imagining I hear grunts and clicks coming from it.

"What've you got in the box?" I ask Jack.

"A coyote," he answers, without even looking back. Eyes on the road.

"No shit," I say, happy that he trusts me enough, already, to bullshit with me. "Where'd you get it?"

Jack doesn't answer, and I can tell that Kirby thinks we, Jack and I, are playing some sort of joke on him, one that he refuses to pick up on, and so the subject is dropped. But I can still hear something in that box behind me, something alive, moving around from time to time, occasionally making what sounds like spitting noises.

Near Galveston the night ends, and the flags on buildings are snapping straight out to the northwest, toward Montana, from where I've come. There's a warm southeast wind, which is the best for fishing, and though we are still in the squall, Jack appears crazed by this good omen, rolling his window down, despite the rain, to smell it. He believes there is the tiniest chance that it will be raining on land but not out in the bay, that it will be wet and storming in one place, but dry and breezy at that place's edge. As we start driving past the refineries, past the tidal inlets,

Kirby and I begin to believe in his wild hope; we have only a few miles to go, but it's true, the rain has slowed to a drizzle.

"We did it," Jack says, delighted. "We fucking outran it."

There's no one else out on the bay, though soon others will begin showing up, diluting the space with their presence. For now it is just us, and we get out and watch as Jack backs the trailer down the boat ramp into the water. Colonies of barnacles cling to nearby pilings, and there's a little bait shop at the end of the dock, which is not open yet because the weather has been so bad, even though it's early light, dawn, fishing time.

Towering above our launch spot is a huge billboard with the photograph of a dark-haired woman on it, perhaps the most beautiful woman we've ever seen; it's one of those "Wanted: Missing Person" advertisements. She's smiling on the billboard, and way above us like that, up in the windy, cloud-parting sky, she looks like a goddess, granting us permission to go fishing, to go out and play.

"HELP FIND RENEE JACKSON," the billboard says, and I study the woman closely as I try to remember if I have ever seen her before, and then I think how the name sounds familiar. Perhaps she is someone I went to school with. But that is too long ago, it is old pork, stored in salt, gaps in memory, and there is only the future. I would like to help if I could — I would like to lift that bull too — but it's all I can do to hope that Renee is all right, to give her my earnest, best hope. It is not a good feeling.

We're lowering the boat into the water with a winch, the *click-click-click* of the wire cable spooling out. Kirby's operating it. He's been on a hundred fishing trips with Jack. I get the sack lunches that Tricia, Kirby's wife, packed for all of us, and carry them down the dock and hand them to Jack, who is already in the boat arranging things, checking the fuel tanks and such.

"Tricia make these?" Jack calls up to Kirby, who's about to pull the jeep and trailer away, to go park.

"Yep," says Kirby.

"She's so sweet," says Jack.

"Would Wendy make a lunch like that for you?" Kirby asks.

"Hell," says Jack, looking through the lunch as if it's a discovery, "she didn't even get up to say goodbye."

Kirby is beaming, as if he's gotten away with something.

Kirby and I climb into the boat. The first thing I notice is that there aren't any life vests, and I can't swim, but I'm not worried. I'm not going beyond the bay. I look above me at Renee Jackson, the most beautiful woman I've ever seen, and it seems, this early in the morning, having driven through the night and the rain to get here, a blessing to launch ourselves beneath her gaze. The wind whips at my windbreaker, and I feel my eyes beginning to blur and go to water.

"Hey, are you crying, man?" Jack asks, and I wonder if Kirby has told him about Margie and me separating. "It's okay if you are, man," he says. The engine has finally caught and is sputtering clouds of blue smoke out over the bay, giving off the summer-sweet smell of outboard fuel. Jack the dentist is another man now, down low in the boat, working the throttle, turning the wheel with one hand. He's suddenly an outlaw too, the happiest one, and I think that's how it always goes, how the longer you go without something, the happier you are when you finally get it. I think about how happy Renee Jackson's parents would be if she were to show up at their front door today.

"I mean, it's all right," Jack says again, squinting at me

in the weak light. Every raggedy cloud is fleeing, burning flame red above us as the sky begins to light up, though down here on the water it's still dusky and gloomy, still foggy gray. "Kirby cried for half an hour after we lost a redfish last year," Jack says. "I don't mean lost it on a hook — I mean lost it, dead. She was a forty-five-pound female with eggs, and it took so long to land her that she wasn't any good by the time we got her in. We tried to let her go again right away, but she just lay there in the surf, gasping, and then rolled over on her side. We worked with her for two hours before she died. For a while we thought she was going to make it," Jack says. "She was as big as a dog. Two hours. What else could we do but cry?"

I have to turn away from the picture of Renee Jackson or I *will* cry.

"Hand me one of those Rolling Rocks," I tell Kirby.

"Running like a fine watch!" Jack shouts, revving the engine.

An alarming *thump* shakes the back of the boat, where the motor is housed, followed by an even more alarming miscellany of piston noise and exhaust. Slowly we creep into the bay, following the lane of driven cedar piles into deeper water.

We run around in large circles before entering open water, to iron out the engine's kinks before we get too far from shore. Sure enough, the engine cuts out, just as the sun is completely up, brilliant and golden in our eyes, and the strong salt wind is in our faces.

We sit like fools for a while, too far from shore to wade or swim back — the water is four to six feet deep throughout the bay, but the current is strong. Kirby and I, out of old habit, begin to despair and open bottles of beer. Jack, though, is still riding the crest of being captain, and the change in him is still evident, even from the set of his jaw. He lifts the cover of the engine and spies the problem immediately: the wire leading from one of the spark plugs is bare and wet from the storm, and has shorted out. Jack has some electrical tape in his toolbox, and he wraps the offending wire quickly.

I don't mean to make Jack look like such a genius. The reason he was able to go straight to the problem is that he and Kirby had taken Jack's seventy-nine-year-old father out in the boat the week before, and the old man, a perfectionist, had ranted and raved for the first hour of the trip about the terrible condition Jack had let the boat fall into. Evidently the boat had belonged to Jack's father fifteen or twenty years ago, and the old man's loopy hearing had picked up on the spark-plug wire's shorting right away.

"He was really howling," Kirby says of Jack's father. "Man, is he a hardass."

Chastened by the memory, Captain Jack slouches a little lower in the seat. Something is troubling him now; his face looks like it did when he was driving through the rain.

"Dollar-bill green!" he shouts, looking down at the water we're skimming across. I'm sitting up in the high-perched bow like a mascot, sniffing the sea. "When the water's this color and the wind's out of the southeast," Jack says, "you'll catch fish."

Kirby moves up to the bow with me, still drinking his beer, and tries to fill me in as quickly as possible on all the things I should know, all the things he and Jack have learned from fishing together for the last five or six years.

"There's dolphins out here, but you never catch them," Kirby says. "They only follow you. Sometimes they come right up to the boat and stick their head out of the water and look you in the eye. A lot of times you can tell where the speckled trout are by the way the seagulls are acting. In warm weather, the summer usually, you can look for slicks. A slick is an oily, flat spot on the water where the fish have gotten into a feeding frenzy on the shrimp and have eaten so much they've regurgitated it, and all the oils and digestive juices make this big slick on the ocean. It smells like watermelon. You smell watermelon out at sea and you'd better be ready."

"We may run aground," Jack shouts from the back. "Be careful." I picture us sliding to an immediate stop, beached by a barely submerged sandbar. I picture myself not stopping but being catapulted out of the boat, a human cannonball, and I sit a little lower in the bow and grip the sides.

It all looks the same to me. I can't see the shore anymore, can't see where any sandbars might be that could cause us to run aground, though I keep watching. We bounce across the chops of waves a little longer — seemingly by whim, with no plan, no landmark, and then Jack cuts the engine, and we're adrift.

The silence sounds wonderful. "Start fishing," Jack says. He is already scrambling like a child, eager to get his lure — an artificial shrimp, blood red ("strawberry") in color — threaded onto a quarter-ounce jig and into the water. It's a big deal, I find out, to catch the first fish of the day.

"We bought Kirby her first pair of shoes this week," Kirby says, once he has his rod set up and is working it, casting, retrieving, casting again. There's high anticipation among us — any one of us could catch the first fish at any given second, even me. "Man, was she mad!" Kirby laughs, remembering. "She kicked and waved, trying to throw them off." He's been a father for seven months.

Jack's silent, intense, almost manic. It's a lovely day. The sun is warm on our shoulders, though just off to our left, where land is, we can see the black squall line — savage thunderstorms, wicked cold streaks of lightning. Also coming from that direction, far in the distance, is a line of boats raising big wakes, bearing down like a posse.

"The popdicks," Kirby grunts.

"You got any popdicks in Montana?" Jack asks, glancing in the direction of the oncoming boats.

"Say what?" I ask.

"Popdicks," Jack says. He's watching his line again, reeling it in. I don't think I've ever seen anyone as serious about anything as Jack is about catching that first fish.

"What's a popdick?" I say. I'm almost afraid to ask. Kirby and Jack howl, delighted to hear me say the word. It's a private joke between them, some word they've made up, and I feel as if I've crossed a magical boundary and been initiated into something important. Suddenly I feel farther away than ever from Margie.

"Popdicks," Kirby says, "run their boats across the water in front of you, going too fast, and they scare all the fish away."

Just then Jack's rod bows. He's got a big one, the first saltwater fish I've ever seen caught. Only I haven't seen it yet. It's still out there in the bay, fighting to get free. But Jack's bringing it in, and Kirby clambers about the boat trying to get the landing net ready. Presently we see flashes of silver, like underwater lightning, then Kirby has the net under the big fish, a club-length speckled trout, fierce-toothed, metallic gray with a yellow and white belly and smart eyes. Jack quickly unhooks the fish and slides it into the ice chest, where it thrashes and beats its tail against the sides — a sound we pretend not to hear, or rather, understand.

"Easy, big fella," Jack says, readjusting his strawberry shrimp; it's been half pulled off, like a woman coming out of her slip, and he slides it all the way back on the hook. We can't cast out again, though, because by now all the pop-dicks have converged in a rough circle around us. They are casting shamelessly into the school of speckled trout that was ours first. They're catching them and hooting with joy and excitement, as if they've done something special.

"I'd rather be dead than be a popdick," Jack mutters, and gives the old boat its full throttle, gunning us through the center of the schooling trout. Several fish leap out of the water, bright and glittering in the sun, and then we're past the ranks of the popdicks, running again for the open sea.

In the afternoon Jack finds another school, and both he and Kirby hook a trout on the same retrieve. There are a few other boats following roughly the same drift line as we are, but they aren't close enough to see our poles, and if we're careful, we can keep the fish a secret.

"This is how you do it when the popdicks are watching," Jack says, speaking through his teeth like a ventriloquist and holding the rod down low to the water, reeling in nonchalantly, as if nothing were happening. I see that Kirby is doing the same thing. They land the fish quietly, without the net and around the back side of the boat, so that it looks like we are getting beer out of the ice chest.

I'm not getting any strikes. Jack and Kirby try to give me pointers as they fish, but it's hard to fish and teach at the same time. Like most things, it's just something that I am going to have to work out by myself. They each catch two more fish before the popdicks realize what's going on and start their engines and come racing over.

"Hey, you *popdicks!*" Kirby shouts when he sees the secret is out. He holds the bent pole high in the air with one hand. The fish that is on the other end struggles and dives, and with his other hand he begins waving the boats over. "Hey everybody, come on over here! I'm catching 'em! Hey, come on!"

Jack curses, shaking his head, as if disciplining himself to say nothing — I can tell he hates scenes — but Kirby is giggling, and we leave the spot in a full spray just as the first of the boats arrive, friendly and curious, shameless, looking for fish.

After that we go far inland, past the point where anyone can possibly see us. In the shallower water we begin catching hook-jawed flounder and gulf trout. Jack catches a fair-sized redfish — a big, dull-looking, bullheaded fish that keeps its mouth open all the time, and which, before being landed, with its great strength runs around and around the boat, circling it like a shark five or six times before tiring. Kirby has a tube of some magic fish-catching attractant that he wasted three dollars on at a sporting-goods store in Houston. "Kawanee" is the magical name Kirby and Jack have given to it, and after several more beers, and with the ice chest beginning to fill with fish (fish for dinner, fish for the freezer, and fish for their wives), we grow slightly goofy, and Kirby insists on smearing my lure — still a strawberry shrimp, which is what is working for them — with Kawanee before each cast. I still haven't caught a single fish.

There is the requisite talk of sex.

"You know, Jack, when your assistant leans over me in the dentist's chair, I can see her bosoms," Kirby says. "I mean, I can *really* see them, even the tips."

"No shit," Jack says, nursing a beer. There is a slack spot in the fishing, perhaps because we have all put Kawanee on our lures. It is a waxy, greasy substance like Chap Stick, and it smells like something dead. "No shit," Jack says again, perhaps imagining it. Then he says, "Well, that's fine, but you can't look at things like that anymore. You're married. Hell, you're even a father."

"Yeah, but I'm still a wild sonofabitch," Kirby says, and he sounds almost angry.

The wind out of the southeast is warm and salty. It's blowing us toward shore. The tide, too, has turned and is running back in — we can drift all the way to where we came from.

"Wendy's mean," Jack says, picking up another bottle, "but she's a hellcat in bed. Thank goodness."

Kirby just grunts. I can tell he isn't going to bring Tricia into this, and I'm not about to bring Margie into this either. I don't really want to hear about Wendy, don't want to picture her being a hellcat. Maybe someone else, maybe even Jack's dental assistant, but not his wife. I don't want to hear about that, and I don't think Kirby does either.

We drift like that, fishing slowly and drinking beer, with a dark purple cloud bank hanging over the shore.

"I can't get over how upset your old man was about those spark-plug wires," Kirby says. "I thought he was going to blow a clot. Do you know how happy he was to be able to really rip into you? He'll be talking about it for the rest of his life. He'll never stop."

"Aw, that's okay," Jack says, sighing. "I know he's losing his mind, but he's still my dad. I guess I can put up with it for a few more years. We'll be dopey old fuckers ourselves someday."

"I hope so," Kirby says.

We are close enough to shore that I can see the bill-board of Renee Jackson again. We drift toward her lazily.

"She's been missing for a long time, hasn't she?" Kirby asks Jack.

"I think so," Jack says. "But I think they found her. I think that's the one whose skeleton they found over on East Beach last spring."

We have to look at her as we drift in. There is nowhere else to look but straight at her, and she's looking back at us, smiling. It is that point in the day — and always, each day, whether you are two blocks or two continents away, you feel it — when you are too far away from your wife, your family — cut loose, cast off, drifting away — and when you wish so strongly that you could see them again, could reach out and hold their hands. We drift near the boat launch, maneuvering the last several yards with the motor, suddenly tired from our day. Kirby hops out and gets the jeep and trailer, lowers it into the water so we can drive the boat up on it. We hook the cable winch up to it and reel it in, ready to go home.

I know what Margie meant about feeling tired. I am tired too. But we have to keep going on.

It is about four-thirty in the afternoon. Jack stops along a deserted stretch of beach on the way home — splatters of rain beginning to strike our faces, another storm start-ing up — and he asks Kirby and me to help him carry the steel box from the back of the jeep down to the dunes.

I'd forgotten about it, but as soon as we lift it, it becomes apparent that there is something alive in the box after all, something spitting and snarling, and we set it down in the tall salt grass and then step back.

"I trapped it in my back yard," Jack says proudly. "I really did."

The gulf wind stings our faces with salt mist blowing off the waves. But it is a warm wind. The beach, at high tide, is a long, narrow strip of tan. The sky is a lurid purple-black, like the bruise on the inside of a woman's thigh. We can see condominiums and high-rises farther down the beach. Lightning crackles and speaks all around us.

"Let her rip," Jack says, opening the door to the metal box. A small coyote about the size of a collie shoots out without looking back and begins running down the beach in a straight line. It is running with its tail floating behind, running — and this is the most beautiful thing — directly toward the condominiums and townhouses, running north and into the wind, without looking back, as if it knows exactly where it is going.

Days of
Heaven

THEIR PLANS were to develop the valley, and my plans were to stop them. There were just the two of them. The stockbroker, or stock analyst, had hired me as caretaker on his ranch here. He was from New York, a big man who drank too much. His name was Quentin, and he had a protruding belly and a small mustache and looked like a polar bear. The other one, a realtor from Billings, was named Zim. Zim had close-together eyes, pinpoints in his pasty, puffy face, like raisins set in dough. He wore new jeans and a western shirt with silver buttons and a metal belt buckle with a horse on it. In his new cowboy boots he walked in little steps with his toes pointed in.

The feeling I got from Quentin was that he was out here recovering from some kind of breakdown. And Zim — grinning, loose-necked, giggling, pointy-toe walking all the time, looking like an infant who'd just shit his diapers — Zim the predator, had just the piece of Big Sky Quentin needed. I'll go ahead and say it right now so nobody gets the wrong idea: I didn't like Zim.

It was going fast, the Big Sky was, Zim said. All sorts of famous people — celebrities — were vacationing here, moving here. "Brooke Shields," he said. "Rich people. I mean *really* rich people. You could sell them things. Say you owned the little store in this valley, the Mercantile. And say Michael Jackson — well, no, not him — say Kirk Douglas lives ten miles down the road. What's he going to do when he's having a party and realizes he doesn't have enough Dom Perignon? Who's he gonna call? He'll call your store, if you have such a service. Say the bottle costs seventy-five dollars. You'll sell it to him for a hundred. You'll deliver it, you'll drive that ten miles up the road to take it to him, and he'll be glad to pay that extra money."

"Bing-bang-bim-bam!" Zim said, snapping his fingers and rubbing his hands together, his raisin eyes glittering. His mouth was small, round, and pale, like an anus. "You've made twenty-five dollars," he said, and the mouth broke into a grin.

What's twenty-five dollars to a stock analyst? But I saw that Quentin was listening closely.

I've lived on this ranch for four years now. The guy who used to own it before Quentin was a predator too. A rough guy from Australia, he had put his life savings into building this mansion, this fortress, deep in the woods overlooking a big meadow. The mansion is three stories tall, rising into the trees like one of Tarzan's haunts.

The previous owner's name was Beauregard. All over the property he had constructed various outbuildings related to the dismemberment of his quarry: smokehouses with wire screening, to keep the other predators out, and butchering houses complete with long wooden tables, sinks, and high-intensity lamps over the tables for night work. There were even huge windmill-type hoists on the property, which were used to lift the animals — moose, bear, and elk, their heads and necks limp in death — up off the ground so their hides could first be stripped, leaving the meat revealed.

It had been Beauregard's life dream to be a hunting guide. He wanted rich people to pay him for killing a wild creature, one they could drag out of the woods and take home. Beauregard made a go of it for three years, before business went downhill and bad spirits set in and he got divorced. He had to put the place up for sale to make the alimony payments. The divorce settlement would in no way allow either of the parties to live in the mansion — it had to be both parties or none — and that's where I came in: to caretake the place until it was sold. They'd sunk too much money into the mansion to leave it sitting idle out there in the forest, and Beauregard went back east, to Washington, D.C., where he got a job doing something for the CIA — tracking fugitives was my guess, or maybe even killing them. His wife went to California with the kids.

Beauregard had been a mercenary for a while. He said the battles were usually fought at dawn and dusk, so sometimes in the middle of the day he'd been able to get away and go hunting. In the mansion, the dark, noble heads of long-ago beasts from all over the world — elephants, greater Thomson's gazelles, giant oryx — lined the walls of the rooms. There was a giant gleaming sailfish leaping over the headboard of my bed upstairs, and there were woodstoves and fireplaces, but no electricity. This place is so far into the middle of nowhere. After I took the caretaking position, the ex-wife sent postcards saying how much she enjoyed twenty-four-hour electricity and how she'd get up during the night and flick on a light switch, just for the hell of it.

I felt that I was taking advantage of Beauregard, moving into his castle while he slaved away in D.C. But I'm a bit of a killer myself, in some ways, if you get right down to it, and if Beauregard's hard luck was my good luck, well, I tried not to lose any sleep over it.

If anything, I gained sleep over it, especially in the summer. I'd get up kind of late, eight or nine o'clock, and fix breakfast, feed my dogs, then go out on the porch and sit in the rocking chair and look out over the valley or read. Around noon I'd pack a lunch and go for a walk. I'd take the dogs with me, and a book, and we'd start up the trail behind the house, following the creek through the larch and cedar forest to the waterfall. Deer moved quietly through the heavy timber. Pileated woodpeckers banged away on some of the dead trees, going at it like cannons. In that place the sun rarely made it to the ground, stopping instead on all the various levels of leaves. I'd get to the waterfall and swim — so cold! — with the dogs, and then they'd nap in some ferns while I sat on a rock and read some more.

In midafternoon I'd come home — it would be hot then, in the summer. The fields and meadows in front of the ranch smelled of wild strawberries, and I'd stop and pick some. By that time of day it would be too hot to do anything but take a nap, so that's what I'd do, upstairs on the big bed with all the windows open, with a fly buzzing faintly in one of the other rooms, one of the many empty rooms.

When it cooled down enough, around seven or eight

in the evening, I'd wake up and take my fly rod over to the other side of the meadow. A spring creek wandered along the edge of it, and I'd catch a brook trout for supper. I'd keep just one. There were too many fish in the little creek and they were too easy to catch, so after an hour or two I'd get tired of catching them. I'd take the one fish back to the cabin and fry him for supper.

Then I'd have to decide whether to read some more or go for another walk or just sit on the porch with a drink in hand. Usually I chose that last option, and sometimes while I was out on the porch, a great gray owl came flying in from the woods. It was always a thrill to see it — that huge, wild, silent creature soaring over my front yard.

The great gray owl's a strange creature. It's immense, and so shy that it lives only in the oldest of the old-growth forests, among giant trees, as if to match its own great size against them. The owl sits very still for long stretches of time, watching for prey, until — so say the ornithologists — it believes it is invisible. A person or a deer can walk right up to it, and so secure is the bird in its invisibility that it will not move. Even if you're looking straight at it, it's convinced you can't see it.

My job, my only job, was to live in the mansion and keep intruders out. There had been a For Sale sign out front, but I took it down and hid it in the garage the first day.

After a couple of years, Beauregard, the real killer, did sell the property, and was out of the picture. Pointy-toed Zim got his 10 percent, I suppose — 10 percent of $350,000; a third of a million for a place with no electricity! — but Quentin, the stock analyst, didn't buy it right away. He *said* he was going to buy it, within the first five minutes of seeing it. At that time, he took me aside and asked if I could stay on, and like a true predator I said, Hell yes. I didn't care who owned it as long as I got to stay there, as long as the owner lived far away and wasn't someone who would keep mucking up my life with a lot of visits.

Quentin didn't want to live here, or even visit; he just wanted to *own* it. He wanted to buy the place, but first he wanted to toy with Beauregard for a while, to try and drive the price down. He wanted to *flirt* with him, I think.

Myself, I would've been terrified to jack with Beauregard. The man had bullet holes in his arms and legs, and scars from various knife fights; he'd been in foreign prisons and had killed people. A bear had bitten him in the face, on one of his hunts, a bear he'd thought was dead.

Quentin and his consultant to the West, Zim, occasionally came out on "scouting trips" during the summer and fall they were buying the place. They'd show up unannounced with bags of groceries — Cheerios, Pop Tarts, hot dogs, cartons of Marlboros — and want to stay for the weekend, to "get a better feel for the place." I'd have to move my stuff — sleeping bag, frying pan, fishing rod — over to the guest house, which was spacious enough. I didn't mind that; I just didn't like the idea of having them around.

Once, while Quentin and Zim were walking in the woods, I looked inside one of their dumb sacks of groceries to see what they'd brought this time and a magazine fell out, a magazine with a picture of naked men on the cover. I mean, drooping penises and all, and the inside of the magazine was worse, with naked little boys and naked men on motorcycles.

None of the men or boys in the pictures were ever *doing* anything, they were never touching each other, but still

the whole magazine — the part of it I looked at, anyway
— was nothing but heinies and penises.

In my woods!

I'd see the two old boys sitting on the front porch, the
lodge ablaze with light — those sapsuckers running *my*
generator, *my* propane, far into the night, playing *my* Jimmy
Buffett records, singing at the top of their lungs. Then
finally they'd turn the lights off, shut the generator down,
and go to bed.

Except Quentin would stay up a little longer. From the
porch of the guest house at the other end of the meadow (my
pups asleep at my feet), I could see Quentin moving through
the lodge, lighting the gas lanterns, walking like a ghost.
Then the sonofabitch would start having one of his fits.

He'd break things — plates, saucers, lanterns, windows,
my things and Beauregard's things — though I suppose they
were now his things, since the deal was in the works. I'd
listen to the crashing of glass and watch Quentin's big,
whirling polar-bear shape passing from room to room.
Sometimes he had a pistol in his hand (they both carried
nine-millimeter Blackhawks on their hips, like little cow-
boys), and he'd shoot holes in the ceiling and the walls.

I'd get tense there in the dark. This wasn't good for my
peace of mind. My days of heaven — I'd gotten used to
them, and I wanted to defend them and protect them,
even if they weren't mine in the first place, even if I'd never
owned them.

Then, in that low lamplight, I'd see Zim enter the room.
Like an old queen, he'd put his arm around Quentin's big
shoulders and lead him away to bed.

After one of their scouting trips the house stank of cigarettes, and I wouldn't sleep in the bed for weeks, for fear

of germs; I'd sleep in one of the many guest rooms. Once I found some mouthwash spray under the bed and pictured the two of them lying there, spraying it into each other's mouths in the morning, before kissing . . .

I'm talking like a homophobe here. I don't think it's that at all. I think it was just that realtor. He was just turning a trick, was all.

I felt sorry for Quentin. It was strange how shy he was, how he always tried to cover up his destruction, smearing wood putty into the bullet holes and mopping the food off the ceiling — this fractured stock analyst doing domestic work. He offered me lame excuses the next day about the broken glass — "I was shooting at a bat," he'd say, "a bat came in the window" — and all the while Zim would be sitting on my porch, looking out at my valley with his boots propped up on the railing and smoking the cigarettes that would not kill him quick enough.

Once, in the middle of the day, as the three of us sat on the porch — Quentin asking me some questions about the valley, about how cold it got in the winter — we saw a coyote and her three pups go trotting across the meadow. Zim jumped up, seized a stick of firewood (*my* firewood!), and ran, in his dirty-diaper waddle, out into the field after them, waving the club like a madman. The mother coyote got two of the pups by the scruff and ran with them into the trees, but Zim got the third one, and stood over it, pounding, in the hot midday sun.

It's an old story, but it was a new one for me — how narrow the boundary is between invisibility and collusion. If you don't stop something yourself, if you don't single-handedly step up and change things, then aren't you just as guilty?

I didn't say anything, not even when Zim came huffing

back up to the porch, walking like a man who had just gone out to get the morning paper. There was blood speckled around the cuffs of his pants, and even then I said nothing. I did not want to lose my job. My love for this valley had me trapped.

We all three sat there like everything was the same — Zim breathing a bit more heavily, was all — and I thought I would be able to keep my allegiances secret, through my silence. But they knew whose side I was on. It had been *revealed* to them. It was as if they had infrared vision, as if they could see everywhere, and everything.

"Coyotes eat baby deer and livestock," said the raisin-eyed sonofabitch. "Remember," he said, addressing my silence, "it's not your ranch anymore. All you do is live here and keep the pipes from freezing." Zim glanced over at his soul mate. I thought how when Quentin had another crackup and lost this place, Zim would get the 10 percent again, and again and again each time.

Quentin's face was hard to read; I couldn't tell if he was angry with Zim or not. Everything about Quentin seemed hidden at that moment. How did they do it? How could the bastards be so good at camouflaging themselves when they had to?

I wanted to trick them. I wanted to hide and see them reveal their hearts. I wanted to watch them when they did not know I was watching, and see how they really were — beyond the fear and anger. I wanted to see what was at the bottom of their black fucking hearts.

Now Quentin blinked and turned calmly, still revealing no emotion, and gave his pronouncement. "If the coyotes eat the little deers, they should go," he said. "Hunters should be the only thing out here getting the little deers."

• • •

The woods felt the same when I went for my walks each time the two old boys departed. Yellow tanagers still flitted through the trees, flashing blazes of gold. Ravens quorked as they passed through the dark woods, as if to reassure me that they were still on my side, that I was still with nature, rather than without.

I slept late. I read. I hiked, I fished in the evenings. I saw the most spectacular sights. Northern lights kept me up until four in the morning some nights, coiling in red and green spirals across the sky, exploding in iridescent furls and banners. The northern lights never displayed themselves while the killers were there, and for that I was glad.

In the late mornings and early afternoons, I'd sit by the waterfall and eat my peanut butter and jelly sandwiches. I'd see the same magic sights: bull moose, their shovel antlers in velvet, stepping over fallen, rotting logs; calypso orchids sprouting along the trail, glistening and nodding. But it felt, too, as if the woods were a vessel, filling up with some substance of which the woods could hold only so much, and when the forest had absorbed all it could, when no more could be held, things would change.

Zim and Quentin came out only two or three times a year, for two or three days at a time. The rest of the time, heaven was mine, all those days of heaven. You wouldn't think they could hurt anything, visiting so infrequently. How little does it take to change — spoil — another thing? I'll tell you what I think: the cleaner and emptier a place is, the less it can take. It's like some crazy kind of paradox.

After a while, Zim came up with the idea of bulldozing the meadow across the way and building a lake, with sailboats and docks. He hooked Quentin into a deal with a log-house manufacturer in the southern part of the state who was going to put shiny new "El Supremo" homes around the lake. Zim was going to build a small hydro dam on the creek and bring electricity into the valley, which would automatically double real estate values, he said. He was going to run cattle in the woods, lots of cattle, and set up a little gold mining operation over on the north face of Mount Henry. The two boys had folders and folders of ideas. They just needed a little investment capital, they said.

It seemed there was nothing I could do. Anything short of killing Zim and Quentin would be a token act, a mere symbol. Before I figured that out, I sacrificed a tree, chopped down a big, wind-leaning larch so that it fell on top of the lodge, doing great damage while Zim and Quentin were upstairs. I wanted to show them what a money sink the ranch was and how dangerous it could be. I told them how beavers, forest beavers, had chewed down the tree, which had missed landing in their bedroom by only a few feet.

I know now that those razor-bastards knew everything. They could sense that I'd cut that tree, but for some reason they pretended to go along with my story. Quentin had me spend two days sawing the tree for firewood. "You're a good woodcutter," he said when I had the tree all sawed up and stacked. "I'll bet that's the thing you do best."

Before he could get the carpenters out to repair the damage to the lodge, a hard rain blew in and soaked some of my books. I figured there was nothing I could do. Anything I did to harm the land or their property would harm me.

Meanwhile the valley flowered. Summer stretched and yawned, and then it was gone. Quentin brought his children out early the second fall. Zim didn't make the trip, nor did I spy any of the skin magazines. The kids, two girls and a boy who was a younger version of Quentin, were okay for a day or two (the girls ran the generator and watched movies on the VCR the whole day long), but little Quentin was going to be trouble, I could tell. The first words out of his mouth when he arrived were "Can you shoot anything right now? Rabbits? Marmots?"

And sure enough, before two days went by he discovered that there were fish — delicate brook trout with polkadotted, flashy, colorful sides and intelligent-looking goldrimmed eyes — spawning on gravel beds in the shallow creek that ran through the meadow. What Quentin's son did after discovering the fish was to borrow his dad's shotgun and begin shooting them.

Little Quentin loaded, blasted away, reloaded. It was a pump-action twelve-gauge, like the ones used in big-city detective movies, and the motion was like masturbating — *jack-jack boom, jack-jack boom.* Little Quentin's sisters came running out, rolled up their pant legs, and waded into the stream.

Quentin sat on the porch with drink in hand and watched, smiling.

During the first week of November, while out walking — the skies frosty, flirting with snow — I heard ravens, and then noticed the smell of a new kill, and moved over in that direction.

The ravens took flight into the trees as I approached. Soon I saw the huge shape of what they'd been feasting on: a carcass of such immensity that I paused, frightened, even though it was obviously dead.

Actually it was two carcasses, bull moose, their antlers locked together from rut-combat. The rut had been over for a month, I knew, and I guessed they'd been attached like that for at least that long. One moose was long dead —

two weeks? — but the other moose, though also dead, still had all his hide on him and wasn't even stiff. The ravens and coyotes had already done a pretty good job on the first moose, stripping what they could from him. His partner, his enemy, had thrashed and flailed about, I could tell — small trees and brush were leveled all around them — and I could see the swath, the direction from which they had come, floundering, fighting, to this final resting spot.

I went and borrowed a neighbor's draft horse. The moose that had just died wasn't so heavy — he'd lost a lot of weight during the month he'd been tied up with the other moose — and the other one was a ship of bones, mostly air.

Their antlers seemed to be welded together. I tied a rope around the newly dead moose's hind legs and got the horse to drag the cargo down through the forest and out into the front yard. I walked next to the horse, soothing him as he pulled his strange load. Ravens flew behind us, cawing at this theft. Some of them filtered down from the trees and landed on top of the newly dead moose's humped back and rode along, pecking at the hide, trying to find an opening. But the hide was too thick — they'd have to wait for the coyotes to open it — so they rode with me, like gypsies: I, the draft horse, the ravens, and the two dead moose moved like a giant serpent, snaking our way through the trees.

I hid the carcasses at the edge of the woods and then, on the other side of a small clearing, built a blind of branches and leaves where I could hide and watch over them.

I painted my face camouflage green and brown, settled into my blind, and waited.

The next day, like buffalo wolves from out of the mist, Quentin and Zim reappeared. I'd hidden my truck a couple of miles away and locked up the guest house so they'd think I was gone. I wanted to watch without being seen. I wanted to see them in the wild.

"What the shit!" Zim cried as he got out of his mongo-tire jeep, the one with the electric winch, electric windows, electric sunroof, and electric cattle prod. Ravens were swarming my trap, gorging, and coyotes darted in and out, tearing at that one moose's hide, trying to peel it back and reveal new flesh.

"Shitfire!" Zim cried, trotting across the yard. He hopped the buck-and-rail fence, his flabby ass caught momentarily astraddle the high bar. He ran into the woods, shooing away the ravens and coyotes. The ravens screamed and rose into the sky as if caught in a huge tornado, as if summoned. Some of the bolder ones descended and made passes at Zim's head, but he waved them away and shouted "Shitfire!" again. He approached, examined the newly dead moose, and said, "This meat's still good!"

That night Zim and Quentin worked by lantern, busy with butchering and skinning knives, hacking at the flesh with hatchets. I stayed in the bushes and watched. The hatchets made whacks when they hit flesh, and cracking sounds when they hit bone. I could hear the two men laughing. Zim reached over and smeared blood delicately on Quentin's cheeks, applying it like makeup, or medicine of some sort, and they paused, catching their breath from their mad chopping before going back to work. They ripped and sawed slabs of meat from the carcass and hooted, cheering each time they pulled off a leg.

They dragged the meat over the autumn-dead grass to the smokehouse, and cut off the head and antlers last, right before daylight.

I hiked out and got my truck, washed my face in a stream, and drove home.

They waved when they saw me come driving in. They were out on the porch having breakfast, all clean and freshly scrubbed. As I approached, I heard them talking as they always did, as normal as pie.

Zim was lecturing to Quentin, waving his arm at the meadow and preaching the catechism of development. "You could have a nice hunting lodge, send 'em all out into the woods on horses, with a yellow slicker and a gun. *Boom!* They're living the western experience. Then in the winter you could run just a regular guest lodge, like on *Newhart.* Make 'em pay for everything. They want to go cross-country skiing? Rent 'em. They want to race snowmobiles? Rent 'em. Charge 'em for taking a *piss.* Rich people don't mind."

I was just hanging back, shaky with anger. They finished their breakfast and went inside to plot, or watch VCR movies. I went over to the smokehouse and peered through the dusty windows. Blood dripped from the gleaming red hindquarters. They'd nailed the moose's head, with the antlers, to one of the walls, so that his blue-blind eyes stared down at his own corpse. There was a baseball cap perched on his antlers and a cigar stuck between his big lips.

I went up into the woods to cool off, but I knew I'd go back. I liked the job of caretaker, liked living at the edge of that meadow.

That evening, the three of us were out on the porch watching the end of the day come in. The days were getting shorter. Quentin and Zim were still pretending that none of the previous night's savagery had happened. It occurred to me that if they thought I had the power to stop them, they would have put my head in that smokehouse a long time ago.

Quentin, looking especially burned out, was slouched down in his chair. He had his back to the wall, bottle of rum in hand, and was gazing at the meadow, where his lake and his cabins with lights burning in each of them would someday sit. I was only hanging around to see what was what and to try to slow them down — to talk about those hard winters whenever I got the chance, and mention how unfriendly the people in the valley were. Which was true, but it was hard to convince Quentin of this, because every time he showed up, they got friendly.

"I'd like that a lot," Quentin said, his speech slurred. Earlier in the day I'd seen a coyote, or possibly a wolf, trot across the meadow alone, but I didn't point it out to anyone. Now, perched in the shadows on a falling-down fence, I saw the great gray owl, watching us, and I didn't point him out either. He'd come gliding in like a plane, ghostly gray, with his four-foot wingspan. I didn't know how they'd missed him. I hadn't seen the owl in a couple of weeks, and I'd been worried, but now I was uneasy that he was back, knowing that it would be nothing for a man like Zim to walk up to that owl with his cowboy pistol and put a bullet, point blank, into the bird's ear — the bird with his eyes set in his face, looking straight at you the way all predators do.

"I'd like that so much," Quentin said again — meaning Zim's idea of the lodge as a winter resort. He was wearing a gold chain around his neck with a little gold pistol dangling from it. He'd have to get rid of that necklace if he moved out here. It looked like something he might have gotten from a Cracker Jack box, but was doubtless real gold.

"It may sound corny," Quentin said, "but if I owned this valley, I'd let people from New York, from California, from wherever, come out here for Christmas and New Year's. I'd put a big sixty-foot Christmas tree in the middle of the road up by the Mercantile and the saloon, and string it with lights, and we'd all ride up there in a sleigh, Christmas Eve and New Year's Eve, and we'd sing carols, you know? It would be real small town and homey," he said. "Maybe corny, but that's what I'd do."

Zim nodded. "There's lonely people who would pay through the nose for something like that," he said.

We watched the dusk glide in over the meadow, cooling things off, blanketing the field's dull warmth. Mist rose from the field.

Quentin and Zim were waiting for money, and Quentin, especially, was still waiting for his nerves to calm. He'd owned the ranch for a full cycle of seasons, and still he wasn't well.

A little something — peace? — would do him good. I could see that Christmas tree all lit up. I could feel that sense of community, of new beginnings.

I wouldn't go to such a festivity. I'd stay back in the woods like the great gray owl. But I could see the attraction, could see Quentin's need for peace, how he had to have a place to start anew — though soon enough, I knew, he would keep on taking his percentage from that newness. Taking too much.

Around midnight, I knew, he'd start smashing things, and I couldn't blame him. Of course he wanted to come to the woods, too.

I didn't know if the woods would have him.

All I could do was wait. I sat very still, like that owl, and thought about where I could go next, after this place was gone. Maybe, I thought, if I sit very still, they will just go away.

In the Loyal Mountains

MY GIRLFRIEND AND I drove my uncle around the Texas hill country during what was to be the last year of his life. We did not know then that they were his last days — though he did, I think — and we always had a good time. I'm married now, and this girl we drove around with, Spanda, is not my wife, and I was never fooled into believing that one day she might be. All this happened a long time ago; I have been saying it's been ten years for so long that by now it is truthfully more like twenty.

Uncle Zorey was single and had never been married, never had children. It's possible that he spoiled me. Zorey owned a machine shop, and custom manufactured large cranes and bulldozers, and he always had money, unbelievable amounts of it. My father and mother used to laugh about it, because he never seemed to know how to spend it. My father was a professional golfer of sorts. He was thirty-eight years old then, and still trying to make it on the big circuit, the tournament circuit, and he and my mother traveled a good bit. I was their only son, and I stayed with Zorey when they went on the road. My father and mother were very much in love, and loved to travel. There was not the least bit of resentment from them that Zorey was so rich, while we were not. The fact that they enjoyed his company so much was one of my favorite things about him.

I had been born with one leg a few inches shorter than the other — a cruel joke, because it threw my golf swing way off — and understandably, my father gave up on my chances of becoming a pro by the time I was seven or eight. He had the grace, perhaps given to him by sport, not to push me. He had compassion for people weaker than himself. Although my father loved golf, he was a better person than he was a golfer. But he was still a very good golfer, just not among the best, and he'd won or finished high often enough to raise me and to support my mother.

My father had, and still does have, a bad back, and I remember that my mother was always rubbing it. After a match, when my father limped home, she massaged it with a rolling pin. In school I was teased about my father — this was in Texas, back in the sixties. It was widely believed that golf was a sissy's game, with the manicured greens, the caddies, the little electric golf carts, and the natty way of dressing. For a while I tried to convince the other kids in school that my father's nickname was Mad Dog, but it never caught.

When he was home, my father walked around the house with a plastic jug of aspirin in the pocket of his robe, and he was always opening the jug and shaking a few out, swallowing them dry. He wouldn't let himself take anything stronger. I would never ask him how he was feeling — I thought it might remind him of his back pain, if he had somehow managed to put it out of his mind.

· · ·

I remember Uncle Zorey coming over to our house for dinner whenever my parents got back from one of their road trips. It would be a feast, a kind of reward for his watching after me. Mom would do all the cooking, but because Uncle Zorey was an outdoorsman, and because he liked wild game best, he would bring over the food: pheasants and grouse from hunting trips he'd gone on in South Dakota, and venison roasts, and fresh fish he'd caught from one of the many lakes north of Houston. My uncle was a pilot, and often flew by himself to one of these lakes, landing at a grass airstrip outside a tiny backwoods town. He'd give one of the local men a hundred dollars or so for the use of a boat and go out on the lake and catch fish. Sometimes the local man asked to go with him, but my uncle always wanted to be alone. He was a good fisherman, and a good shot. His freezer was always full of fish and game.

At dinner we talked about my school, my father's golf game, or my uncle's recent fishing trips. We also talked about things my mother was interested in — politics, wars, morals — and about her childhood, which she missed. My mother came from a large family, and grew up on a farm in Missouri. She loved talking about that farm and about the things her brothers and sisters used to do — there were nine of them in all — and the trouble they used to get into. She talked about cold mornings, about doing the laundry by hand, and about what a thrill it was to get new shoes — all the old things. She was reliving her history, and we listened to her with awe.

She was the one who killed the chickens on Sundays for the big dinner after church. One of her older sisters would hold the chicken down on a tree stump, setting its neck between two nails driven into the stump, and my mother would hit the neck with the hatchet.

"One-Chop," my mother said. "They called me One-Chop."

We ate so much while listening to her stories. We stayed up late and ate and drank far into the night, as if trying to gain ground on some ache or loneliness that had slipped in while my parents were away. They let me drink, too; I was sixteen or seventeen by then. Zorey ate the most and drank the most. No matter how much food my mother fixed, no matter how many bottles of wine were opened, we finished everything, with Zorey leading the way. At midnight or one in the morning, we'd all be groggy and full, and stumble off to bed.

"Zorey, you were insatiable," my father would say as he and my mother went down the hall, leaning against each other, to their bedroom. My father would look back over his shoulder and say, "Zorey, you were just an animal!" It was a joke they'd had between them for many years, a joke that began with their father. Even then, I had heard, Zorey had an enormous appetite and a brute strength; their father's nickname for him was "Animal."

"Good night, Jackie," my uncle would say to me, pausing at the doorway of the guest room. "Good night all."

I remember stumbling into my room, reeling drunk, pretending I was a gut-shot actor in a western, spinning in the dark, pretending I had caught a bullet in the stomach. Clutching it with both hands, I would do a slow triple spin, all for Hollywood, and topple to the bed, land on my back, and fall instantly asleep.

I am a plain man. What I do for a living has little to do with the way I sometimes feel about things. I'm an accountant, and a junior one at that. I'd like to be someone with power, sweeping power, the power to change things, to right wrongs — a judge, a lawyer, a surgeon — but because I am not any of these things doesn't mean they aren't in me.

My uncle was a crook. His death was a suicide, and it came when he felt the evidence closing in. There were questions arising from (where else?) the construction companies' accounting departments; there were letters and queries from lawyers, polite at first. All of these things are in his dusty files, his long-ago files, which I felt the need to remove from his house after his death, and which I now keep in my attic. It must have been a very tough time for him near the end, with no way out. I wish that he'd never been found out, that he could have gone on forever. There was no fishing in prison, is what must have been on his mind, no woods in which to hunt, no grass airstrips to float down onto on hot June weekday afternoons.

What could he have been thinking? It is not right for me to try to guess. But it is fair for me to remember.

My mother might have thought it was a burden for my uncle to keep me while they were on the road. I don't think she ever realized what fun we had — my uncle and I, and then, once I hooked up with her, Spanda.

Spanda came from the wrong side of the tracks — although in Houston, at that time, there really were no tracks, literally or figuratively. She did not attend my school, and she sometimes didn't even attend her own. Spanda was not a nice girl. I thought she was lovely — and she *was* lovely — but she was a little rough, a little mean, and she did not have many odds in her favor except to be rough and mean.

My leg excited her, the shorter one. It's not proper or relevant to go into how much it excited her; it was her business and mine. But it did — she loved the leg — and though she did not love me, it was the first time I had ever felt such a thing, someone *attracted* to my leg, and to me, and it gave me a confidence I needed badly. It didn't hurt, either, that Uncle Zorey was almost always around, his pockets bulging with dollars, like a caricature of an old-style Texan, the kind people used to love until they learned to make fun of him — generous, big-hearted, with loose money spilling from him like water. Uncle Zorey liked Spanda too, and he saw to it that she always got what she wanted when she was with me, saw to it that she was always happy.

We were both seventeen. This was clover for me. I believed in things rather than understanding them. What we are talking about here is innocence, no different from anyone else's.

Uncle Zorey was as wild as a big kid when he was away from my parents and away from his office. When I got in from school, those times I stayed at his house — which was many times that year, because my parents were traveling all the time — my uncle would change out of his suit and into a pair of old coveralls, go out to the driving range, and hit golf balls.

The driving range would be nearly empty on those afternoons, and my uncle and I were able to practice our swings in peace. There might be a woman or two — matronly, yellow-haired women, overweight, dressed in tight bermudas, with meat-eating spikes on their shoes, to better grip the earth for the long drives that seemed to give them so much pleasure. I was used to their looks — looks of pity, and what they thought was knowledge — and it was easy to ignore them.

Even hitting one-legged, as it were, even hitting off balance, I had my uncle's great, strange strength. After several weeks of practice, I was hitting the ball farther than we could have hoped for. But I could not hit it straight. With my twisting swing, I sent the ball into a wild, sail-away slice, or almost as bad, into a horrid, rocketing hook.

My uncle would sit on a soda crate, sweating, toweling his face with a handkerchief and drinking beers, which he kept in a little ice chest by his side. I would swing harder and harder, but along with my uncle's strength, I had my father's back. At times it hurt so much I wanted to tell my uncle that I didn't want to play golf anymore. But then I'd see his look of childlike expectation as he sat there on the wooden box and studied my swing, and so I took my best cut, and away the ball would soar. Sometimes I got so frustrated that I would shout as loudly as I could — at the frustration, and also at the cramps in my back — and the lady golfers would move away from us, pack up their clubs and leave. My uncle liked the shouts, and he would nod, take a sip of beer, and lean forward and hand me another bucket of golf balls.

My father was having a very good spring. He won one big local tournament, and for the first time in several years was selected to play in a prestigious tournament overseas. He was getting offers again to do endorsements, but he wisely rejected them and concentrated on his golf, and did even better.

He was often written about in the sports pages of the papers, and I was proud of him, but also felt a little guilty from all the days in school at a younger age when I tried to change his name and wished he'd competed in a sport more violent, more bloodthirsty than golf. The newspapers were always saying what a gentleman he was, what a good sport, and how he brought class to the game, class to the city.

I began to eat aspirin the way he did. My uncle never saw me doing it, but I started not long after my father won his tournament. My uncle and I stopped golfing around this time, and I felt a flood of relief. Though I may be attributing too much scheming to Zorey, I wonder now if he knew all along what he'd been doing — filling me with all that golf — eliminating all doubt, all question of what was and wasn't possible. I was delighted never to have to pick up a golf club again, delighted never to have to watch the game being played again.

Uncle Zorey brought Spanda home from work with him one day later that spring, saying simply that her father worked in his plant and that she was new in town, and didn't know anyone. She wanted to meet someone her own age, and so my uncle had volunteered me for the job. He hoped I didn't mind.

Lies! Many children are wise at the age of seventeen, but I was not one of them. I believed my uncle, as did so many other people. Spanda looked like an Indian, with dark eyes and long black hair. Often she wore faded blue jeans and a purple tie-dyed shirt. She never put on makeup. We got along famously from the start. It doesn't matter what I think now — wondering whether we would have gotten along so well were it not for my leg, and more importantly, for my uncle, and his money.

We played cards and listened to the radio; we went for drives with my uncle, who took us along in his truck. At night Spanda came to my room downstairs and slipped into bed with me. My uncle slept upstairs, and slept heavily. He got up only after I had left for school.

Spanda was angry at a lot of things. She had a wonderful vocabulary of curse words, which she used against any and all incarnations of the establishment: traffic lights, policemen, rainy weather. But she was never angry at me or my uncle. I felt like a hero. And I think that upstairs in his bed, as he drifted into sleep, perhaps imagining things, I think that my uncle, too, probably felt like a hero — as well he should have, as well he should have.

Into the hill country we'd drive, once summer came. The rough, rocky country there was in no way like the rest of Texas, certainly not like the gentle, windy gulf coast where we lived. We stayed in hotels in the German tourist towns — Fredericksburg, Boerne, New Braunfels — getting separate rooms, one for Spanda and me and one for my uncle. We stopped at beer gardens and sat outside in the shade, drinking cold beer and eating huge amounts of food, my uncle usually ordering one of everything on the menu. We would walk up and down the wide streets of the little towns, window-shopping in the dazzling heat, with hardly anyone else out, the heat far too great, and buy whatever Spanda desired, whatever my uncle saw and wanted: an old sewing machine or a rocking chair in the window of an antiques store, fresh-baked loaves of bread, a gingham dress for Spanda, a walking stick for me. Then we would put more beer in the ice chest in the back of the truck and head for the wild country. We drove up twisting white caliche roads into mountains of cedar and rock and cactus, the heat rising in shimmers and mirages, then sailed down into the small valleys between the hills, rattling across creek bottoms and high-water caution dips, through water-seeking live oaks. We barreled along, my uncle with a beer in his hand, one foot mashed on the accelerator and the other foot propped up and hanging out the window. Spanda and I would drink beers too. She sat in my lap with her arms around me, her hair swirling, her eyes fuzzy and distant, looking out at the countryside.

Roadrunners scurried across the road in front of our mad flight, and browsing herds of little deer leapt away from us in alarm, vaulting gently over barbed-wire fences and disappearing, with flagging white tails, into the thick tangled cedar. Hawks circled overhead, and vultures too. We headed for an obscure range that we knew about, a small chain of mountains in the central part of the state that was not even on the map: the Loyal Mountains.

A stream called Willow Creek flowed through the range, along which were many great boulders and sandbars and the huge, shady live oaks. Uncle Zorey drove across someone's pasture, bouncing the truck over rocks and logs, still with his foot hanging out the window, and singing "Red River Valley." We'd drive until we couldn't go any farther, then get out and hike up the canyon, following the creek upstream to a cool spot we knew, where we could picnic, nap among sweet ferns, and go swimming in a pool beneath a small waterfall.

We'd stay until dark, drinking Jim Beam and shooting pistols — my uncle kept several in a tool chest in the truck. He'd lug the chest as though it were full of stolen gold while Spanda and I carried the ice chest. All three of us got unbelievably drunk. Spanda floated on her back in the pool, naked and sun-dappled beneath the canopy of oaks, her long hair floating all around her. Zorey sang and shot at the boulders, and the bullets ricocheted with mean, zinging whines. It was impossible to imagine what my father and mother would have thought, had they been able to see us.

"Yeah I've played the Red River Valley," he would bray, "and sat in the kitchen and cried . . .''

The water was deep beneath the waterfall. Spanda and I would climb up onto the boulders above the pool, both of us naked now, and would practice our best dying-actor, shot-from-the-stagecoach falls while bullets flew around us. My uncle wasn't drunk, so much as just crazy.

Later, too tired to drink any more and too tired to attempt the long drive back to Houston, we would check into a hotel in the nearest town. Uncle Zorey would give us some money for supper and the keys to his truck, then go off to his room, where he sat up in bed with all the lights on, watching television.

Spanda and I were always ravenous after the long afternoon of play, and the drinking. With the alcohol beginning to lose its edge, we'd eat barbecue, then go back to our room as if we were adults, as if we were married — as if we had all of life figured out — and make love on and off through the night, falling into it as if diving from one of those boulders. Now that I am older and have seen more, I realize, sadly, that it was more of a dutifulness with which Spanda moved, there in the dark.

But sometimes I awakened from sleep and she would have the covers pulled back. She'd be sitting up in the bed, looking at my leg. I caught her doing this several times, and at first I was flattered, and felt special. When it happened more and more, it began to trouble me a little. Thoughts would come into my head about Spanda, and where she had come from, but they went away, or I put them away.

In these hotel rooms, right after I made love to Spanda and was about to fall asleep, I could hear the TV next door in my uncle's room. I heard the sound of the news, usually, and then my uncle's voice, talking to the person on the television. He would argue with the weatherman about the forecast, or with the sportscaster about the replay of a close play at the plate.

Once, as I lay there listening, I heard the sportscaster talking about a golf tournament, the one my father was in, and I heard my father's name mentioned. Spanda was already asleep, and I sat up to hear better, but my uncle must have gotten up then and turned the set off. I felt bad not knowing whether my father was winning or losing.

I never asked Spanda over for any of the dinners at my house. I knew better than to even mention her to my parents. "Pass the fucking meat loaf, please," I could hear her saying after a glass or two of wine. I could see her getting up from her chair, coming over and sitting in my lap, and putting her tongue in my ear. These were the thoughts I had then, though I realize now that she probably would have been charming — and so I wish we had invited her, even once, just so she could have heard my mother's stories.

"My sisters and I had a pet Brahma bull named Skippy," my mother said. "Every year we got to ride him to school, on the last day of school. He was so tame, so gentle."

My uncle would listen with exceptional interest — with a hunger.

"In the winter we used to hook him up to a harness and use him to pull people's cars out of the snow, after they slid off into the ditches," my mother said.

My father laughed. "We used to use Zorey," he said. "Zorey lifted up the back ends of cars and trucks that were stuck."

My uncle smiled modestly. "Those days are over now, I guess." He looked down at the table.

"They certainly are," my father agreed. He held his big

hands out in front of him, strong hands. Arthritis was already beginning to set in, and he had only three or four more years of golfing left, of being good at it.

My mother poured us each another glass of wine. It sparkled in the light of the chandelier. We toasted to the future.

My mother and father were not home much that summer. I feel now almost as if I aggravated my uncle's condition, though I know he enjoyed watching me, and influencing me. When he could not travel to the Loyal Mountains, but instead had to stay in Houston for business meetings, he would take us in the evenings to the fanciest restaurants in the city. He'd buy Spanda a new dress and give me money to rent a tuxedo. I'd have the tuxedo shop pin up the cuff of my left leg, and we would go out to dinner and to a baseball game afterward, still wearing our tuxedos, Spanda still in her evening dress.

Some nights we would drive out to the county airport where Zorey kept his plane, out west of town, in the flat rice country. We'd park and walk over to Zorey's small red and white Cessna. He didn't keep it in a hangar, but simply tied it down to eye bolts set in the ground, so the plane seemed more like a tethered animal. He would run his hands over the plane, feeling its smooth surface, the coolness of the metal. A few times he asked if we wanted a ride. I always did, so we'd get in, and he would start the engine. The engine would catch, cough, and roar, and the little lights on the instrument panel would come on, illuminating my uncle's face with an eerie green light. My uncle would be transfixed, serious as never before.

Spanda crouched on the floor in back, terrified but trying not to show it, one hand gripping my uncle's seat, the other clutching my seat, for it was only a two-seat plane and I was the copilot. I think she felt very strongly that this was not part of her duty, and it's possible that this, unlike anything else, was the part she did for love, if indeed there was any. Then we were spinning, turning toward the runway, lumbering across the grass — my uncle paid no attention to the paved taxiway. The moon shone down on the runway, making it look wet and shiny, like the beginning of a newer, finer, more glorious life, something inviting. Then the best part, putting the throttle in all the way.

My uncle was a good pilot, and it was hard to tell when we had left the ground. Once up above the airport, we could see the lights of the city to the east, but always we would bank and turn back into the darkness, away from the disorientation of all those streams of light and the silhouettes of tall buildings. We'd fly above the prairie, circling in the dark over the rice fields. It might seem that he would be a wild pilot, prone to doing loops and barrel rolls and figure eights, whether he had been drinking or not, but up in the air, with everything at stake, he was the picture of calm, the picture of my father, even: responsible, cool, and caring.

The little plane's loud roar forced us to shout to each other whenever we wanted to say something. Mostly, though, we looked back at the city and at the darkness below us. Sometimes, feeling chivalrous, I held Spanda's hand, squeezed it, and she would smile weakly in the greenish light.

We'd land when the gas gauge showed less than half empty. We tied the plane down and listened to the engine tick as it cooled in the night. Farther off, we heard crickets and the sound of the interstate. Driving home, with the radio playing, my uncle would be as calm as a mare, would not sing along with any of the songs, would not backtalk any commercials that came on. He was thinking about something else, and he looked tame, like someone else.

His plant was on the way to his house, and we would often stop there and drive through the chain link gate, putting a card in a slot that automatically opened the gate. My uncle parked in the lot and got out of the truck to look at the frozen steel and iron machines lying silent in the night, big floodlights all around the yard. I did not understand how anyone could do this for a living, did not even understand how such boring-looking steel machines could make money, or be worth anything.

In August, during the worst of the heat, Uncle Zorey discovered that he had gout. He took medications for it, and had to stay off his feet. He got an electric wheelchair so he could continue to work, and that was when Spanda and I began to drive him around. We headed up into the Loyal Mountains every chance we got.

My uncle was a poor patient, and refused to change his diet. He rode in the passenger's seat, with Spanda sitting between us, her legs on either side of the stick shift. She had to be careful not to bump his foot with hers, because it would cause him intense pain. He was in pain anyway, and was trying to suffer it silently, but he was not as good at it as was my father. He bellowed whenever we hit a bump, and would immediately take a swallow of whiskey or shake down a few of the pills he was taking.

He rode with a shotgun in his lap and would shoot at the coveys of quail we often saw huddled along the road, bathing in the dust. If he killed any, he would have me go back and pick them up. He'd clean them in the truck as I headed for the Loyals. Spanda sat grim-lipped, looking straight ahead, out the window. Feathers swirled around the truck as he plucked the birds, putting the entrails and feathers in a brown grocery bag that he had brought for that purpose. The insides of quail smell rotten for some reason, even when freshly killed, and we had to drive with the windows down in order to breathe.

Wherever we stopped for the night, Zorey would cook the quail in his hotel room, over cans of Sterno he'd bought at a convenience store. He skewered the birds on a coat hanger and cooked them for an after-dinner snack, basted with butter and pepper, as he watched the news and Spanda and I thumped around next door.

Mornings in the hill country were, and still are, beautiful — a heavy dew, even in the summer, and the sounds of roosters, and of cattle lowing. As his gout got worse, my uncle could not make it to the picnic spot anymore, so Spanda and I would go by ourselves. My uncle insisted on this, and he stayed in the hotel room with a six-pack of beer. When we returned, he would have stacks of paper spread out over the desk and the bed — papers everywhere, frantic-looking, and an adding machine plugged in, and the bottles of pills, and the Jim Beam half gone. I could tell that the accountants' inquiries and all the other loose ends were troubling my uncle, and I wished there was something I could do to help him. I wished I knew about numbers, knew how to line them up so they all made sense. I wanted to go back up into the Loyal Mountains with him, wanted to pass through the small towns with him and Spanda, and eat, and drink, and drive the back roads.

· · ·

It was raining when the news came that my uncle had shot himself. We were all three at home, my mother, my father, and I. I wanted the news — a phone call — to be taken back, to go away somehow. My father wilted and sat down on the couch. His face was ashen. It was as if he weren't my father anymore, as if he weren't anyone anymore — as if he'd had his identity taken away — and I felt that I had betrayed him somehow.

My father had his head in his hands. "I need to be alone," he said.

"No, you don't," my mother said. She came over and sat down by his side. I did not know what to do or where to go. I stood there and watched them sitting on the couch together.

"We should have seen it coming," my father said, "should have seen it coming like a freight train." He was shaking his head, and my mother's arms were around him.

"No," she said. "No."

After a long while my father's color started to come back. He stood up and said to us, "I have to go identify the body now." I thought what an awful task that would be. It would be like looking at himself, and worse, there would be the guilt.

"Do you want me to come?" my mother asked. She looked over at me. "Do you want us to come?"

"No," my father said. "I will be all right."

My father is retired now, and a grandfather. His back still hurts him, but he is silent about it, as ever. My mother still tells us stories about her family and her childhood, but there's a loneliness, and the stories are not devoured with as much eagerness — can never be devoured with as much eagerness as they once were by my uncle. On my dresser I have a picture of my father from twenty years ago. It is from a golf tournament, and he is wearing the winner's jacket — the victor.

What lies ahead? Sam, my son, is strong, and prone to tempers. Sometimes, knowing the past — and not knowing parts of it, too — I am frightened almost to the point of paralysis. Sam may become some kind of athlete. He is often sweet, but can throw horrible tantrums, or can turn distant and moody. He is only three years old, but when he reminds me of something, I overreact, and it is my wife who has to calm me.

"He is so strong," I say. "Already, he's so strong."

Or when he's crying, and has his fits, and turns his back on us — so cold, as if he doesn't need us! — I panic, and it feels as if there is nothing I can do.

"Hold him," my wife says when Sam grows distant, sometimes for no reason. She's large with our second, coming soon, but she shows no fear, no worry, only a willingness to dive into the future.

She picks Sam up and hugs him tightly, holds him close to her, strokes the side of his face, and smiles at me.

"Hold him like this," she says, rocking him and smiling at me. "Like this."